Meet Mister Sherertz, decorated hero

By Wilt Browning, Guest Columnist,
Winston- Salem Journal, June 30, 2012

I'm now old enough to admit it: I read the obituaries. ... (The Winston-Salem) Journal obits for Saturday, June 30, carried several death notices that tell us that we are truly alive, that we live, and that we care what happens to people whose names we have never known. ... As a matter of fact, Saturday's listings told me that I have more heroes in my life than I can count, and Jack (Herbert Jackson) Sherertz (Dec. 1, 1917 - June 28, 2012) is one of them. ...

I wish that somewhere along the way, our lives had crossed and I had the opportunity to talk with Mister Sherertz. I know I always would have called him "Mister Sherertz." To have called him "Jack" would have seemed disrespectful.

So, I want to tell you about Mister Sherertz, if you'll take a moment away from your crossword puzzle. ...

He graduated from Georgia Tech in 1940. The year of his degree is important. He became a navy ensign on Feb. 1, 1941, when he was 23 years old and drew what must have then seemed a plush assignment: Hawaii. So, there he stood, an officer of the USS Nevada at Pearl Harbor on Dec. 7, 1941, as Japanese planes, sounding like millions of angry hornets, appeared on a clear morning over the mountaintops and began their deadly attack dives. At 7:55 that Sunday morning, the Nevada had stood at anchor just aft of the USS Arizona and so close they seemed almost to touch bow to stern.

The Nevada was among the first to be struck. ...

Mister Sherertz survived.

He became an officer assigned to PT Squadron 1, and for three days, June 3-6, 1942, did his part in defending the Midway Islands in the greatest naval battle in history. ...

Mister Sherertz survived.

On June 6, 1944, Mister Sherertz was assigned to a PT boat then churning its way into the conflagration we know as D-Day. ... This becomes personal to me because somewhere up there above that PT boat was my own uncle Jack, parachuting into bloody battle.

Mister Sherertz and my uncle Jack survived.

In other words, this one man who lived this one life was a participant in what arguably are the three most significant events of World War II. Mister Sherertz became a decorated lieutenant commander, and his dress uniform would have been adorned with, among others, a Bronze Star, Distinguished Service Cross, French Legion of Honor Medal and Pearl Harbor Medal.

Mister Sherertz was a hero. His obit doesn't say that, not in those words. But he was. When I hang my flag this Fourth of July to flutter in the heat of this summer, it'll be there to honor not only my uncle Jack, but especially Jack (Herbert Jackson) Sherertz.

Salute, Mister Sherertz.

Kevin Maurer, co-author of No Easy Day

"A son's tribute to a father told with a deft pen and a keen eye for detail. A story worth reading and more importantly a story that needed to be told."

A HERO AMONG MILLIONS

The Amazing Story of US Naval Officer,
H.J. Sherertz, and His Account of Survival
Through Pearl Harbor, Midway and D-Day

By Robert J. Sherertz

ISBN-10: 0-9970157-0-5
ISBN-13: 978-0-9970157-0-6

Dedication

This book is dedicated to my father for sharing his painful story, to my wife for putting up with my obsession for seven years, to my children who clearly embraced what my father did, and to all war veterans and the price they pay for their service.

Contents

Part III — Final Tributes

Acknowledgments

Frog went a courtin', he did ride, M-hm

Frog went a courtin', he did ride, M-hm

Frog went a courtin', he did ride

Sword and pistol by his side, M-hm

As a small child, my father sang the song above each night at bedtime. Later on, I knew him as a gentle man, one of the gentlest I've known. I never heard him raise his voice, curse, or say a bad word about anyone, and he taught Sunday school every Sunday. There was nothing about him that suggested warrior, or Commander, or even bravery. It was impossible for me to imagine him carrying a sword and pistol, or fighting anyone, let alone a war. Yet fight a war he did.

He survived Japanese bombs on the deck of the USS Nevada, the Battle of Midway, the D-Day invasion, and a broadside battle with a German minesweeper off the coast of Jersey Island. He lost over 81 close friends and shipmates and many other men that he commanded. He experienced repeated wartime horrors. Finally, at age 91, he decided to share his story and the unimaginable pain that would come with it, rather than take it as a secret to his grave.

This book is 95% based on my conversations with Lt. Commander Sherertz or content from World War II artifacts that he saved. The conversations occurred over a one-year period between June 6th 2009 and the following June and represented over 300 hours of question/answer sessions. I attempted to verify each story he told and, if conflicts were found with other historical accounts, which were rare, that content was not included in the book. Where gaps existed they were filled in based on conversations with other PT boaters or historical accounts referenced at the end of the book. Almost all dialog was fictitious, except for a few places where World War II veterans thought they remembered something exactly. I did let several PT boaters and other ex Navy people read the dialog to see if it sounded authentic and adjusted it from their suggestions to make it sound as realistic as possible.

My father would have hated being called a hero and hated having his story called amazing, and for that I apologize to him and to him only. He would have said that there were millions of others who did what he did and they deserve the same recognition, and I agree. But to the hundreds of millions of us who have never been in a war, what my father and others like him did during World War II was heroic.

One very important question that I have been asked repeatedly is whether I believed what my father told me. The answer is yes for several reasons. First, I was able to compare his accounts with accounts from his squadron mates; they were nearly always identical, almost as though seared collectively into their brains. Second, when I could not find a squadron mate or shipmate who could corroborate his stories, I went to first hand accounts by others who were in the same battle and his accounts held up well.

In a similar vane a reviewer of the manuscript commented that "...this account... makes him almost too good to be true. He was a wonderful dancer, an able athlete, a gifted boxer, a Sunday school teacher, an exceptional and much admired leader, a young man to turn to for every need including funeral services." Most of these things were easy to corroborate by talking to different members of our extended family who knew him before the war. I think his leadership capabilities were best supported by his rapid wartime promotions and the fact that his superior, John Bulkeley, ultimately an admiral in the Navy, thought enough of him to want him as XO on his destroyer, the USS *Endicott*.

And finally the biggest reason that I believed what he told me was that my father never bragged about himself and rarely talked about himself at all; he was always more interested in what others had to say. When pressed for details he would act slightly embarrassed and I would have to coax him to get him to talk further.

Not surprisingly, there were gaps in his story that had to be filled, and I attempted to make them as historically correct and true to my father's character as possible. However, the amount of fact checking associated with writing this story was massive, so if any World War II veterans or others knowledgeable about World War II find errors, please let me know and I will correct them. You may contact me at hjsherertz@gmail.com.

The fact that my father was able to conquer his war demons enough to talk about his story at all is the result of many people's efforts to try and help him. I will attempt to point out notable contributions in that regard.

I would like to dedicate the story to all those who fought in World War II and to those who stayed behind, wondering whether they would return. I would especially like to honor all those men who served with my father by listing as many of their names as I could find at the end of the book.

Images in the Book

The largest number of images used in the book were from my dad's World War II photo albums, my images, or a few taken by my son. The second largest group of images used in the book were from the public domain, findable for example, using the photo # and the history.navy.mil website, or using the website addresses provided. Some World War II images helpful to telling the story may have copyright issues. These images were cited in the book but are displayed on the book website only (www.AHeroAmongMillions.com). Since many of the images had to be reduced in resolution for electronic or book publication, I have placed additional book images at maximum resolution on the book website. The same website contains a number of additional images, pertinent to the story, but not in the book.

People providing assistance with this book:

Herbert Jackson "Jack" Sherertz, my father, talked with me hundreds of hours about World War II between the ages of ninety-one and ninety-four, in spite of the great pain it caused. He meticulously saved documents and artifacts throughout his military service (1940-1945). In particular, as commanding officer of Squadron 34 (RON 34), he commissioned a squadron historian and squadron photographers, which resulted in the only detailed history of a PT boat squadron during World War II. When our family is through with all of his archived materials, they will be donated to PT Boats, Inc., or another appropriate World War II archive.

Elizabeth Sherertz, my wife and best friend, put up with my disappearing for long hours, with constant questions of the "how does this sound" variety, provided editorial suggestions, and created a plaque from the family in honor of my father. I could not have finished this book without her support and understanding.

Margaret McCulloch Sherertz, my mother and my dad's first wife, made sure my dad knew how proud she was of his war efforts and she kept trying, unsuccessfully, to get him to talk about the war during the 52 years they were married. She repeatedly attempted to take him to Pearl Harbor and Normandy Beach, and kept trying to get him to go to a PT boat reunion, all to no avail until 1995. In 1995 PT Boats, Inc. had their yearly reunion in Charlotte, NC, where my parents lived. My mother saw the sign on the outside of the reunion site, rebuffed his protests, and dragged him inside. Holding onto his arm, she walked him through the lunch area until suddenly they both heard, "Attention, it's the skipper." They had found a table full of RON 34 men who had served under my father when he was the squadron Commander. My mother managed to keep him there during lunch, but then he became uncomfortable and said his goodbyes and they left.

My mother made one final effort to help him in December 1999. She had terminal cancer and would die a month later, but in spite of how awful she felt, she arranged an interview for my father with the Charlotte Observer to talk about the December 7th attack on Pearl

Harbor. I think my father would have refused except for my mother's condition and her insistence that it was important to her that he did this. It is my belief that her efforts at the end of her life finally started breaking down the barriers to his talking.

Lillian Sherertz, my stepmother and my father's second wife, did everything she could to try and help him talk about the war. She took him to the PT Boats, Inc museum at Battleship Cove Massachusetts, unrelated to a PT Boat reunion, and arranged for him to participate in a Charlotte NC ceremony honoring WWII veterans. I think the thing that my father liked the best was when she sewed all of his service ribbons and medals onto a Navy Blue blazer. Her support at the end of his life helped him talk about many things that I know were extremely difficult.

Eric Sherertz, my son, when he was 15 (1997), convinced my father to talk with him about the war on videotape. He was 79 at the time and his memories were still quite vivid. He talked for over an hour. The tape was lost for 15 years and finally resurfaced at the perfect time. My son also provided significant helpful suggestions about the story, took some of the pictures used in the book and helped me market the book.

Betsy Sherertz, my daughter and a graphic artist, designed the book cover and helped with the book website.

Frank and Mary Sherertz, Jack Sherertz's parents, were responsible for providing a large amount of WWII information about my dad. Frank took information from letters sent by Jack Sherertz during World War II and fed them to the Roanoke newspaper. He also purchased several books about PT boats that came out during World War II, in particular, At Closer Quarters, and Victory at Midway. Most importantly, he obtained Jack Sherertz's original World War II service records from the War Department after Jack Sherertz was discharged. Mary Sherertz saved every clipping from their hometown newspaper that had anything to do with World War II. Both of them were extremely proud of his war efforts.

Marine Major Carl Sherertz, Jack Sherertz's cousin and life long friend, arranged for my dad to give a videotaped interview with a Library of Congress Intern talking about his war experiences. It provided a number of clear descriptions when his memories were still quite fresh. This interview was very helpful to corroborate what I was hearing years later.

Most importantly he invited Jack Sherertz, Larry Sherertz, and myself to accompany him to the 65th D-Day Reunion at Bedford Virginia in 2009. I am certain that Jack Sherertz would not have talked to me in detail about the war without this cathartic experience.

John Howard, my nephew, made useful editorial suggestions after reading the first draft. **Ross Anthony**, my nephew, scanned most of Lt. Commander Jack Sherertz's World War II pictures and newspaper clippings into digital files. **David Minichan**, a relative from Roanoke, helped research where my father did cliff diving as a boy.

PT Boaters. Lieutenant Russ Schuster, executive officer (XO) of PT 503, What Next, in Squadron 34, coauthored the RON 34 squadron history and provided information on the encounter of PT- 509 with the German mine sweeper. He was driving PT- 503 and standing right next to Lt. Commander Sherertz during the entire PT- 509 rescue attempt, so his perspective was very similar to that of my father's. He spent hours on the phone answering questions trying to confirm things my father told me. I had the honor of meeting him at the last two PT Boat reunions (2009, 2010). **GM3c Sheldon "Bos" Bosley**, was a gunner's mate on PT- 507, the Heminway Hotel, the RON 34 squadron historian, and coauthored the Squadron 34 History. I had the honor of meeting him and his wife at their house and he also answered many questions via telephone. I am extremely appreciative of all the help he provided. His daughter put me in touch with John Ovenden on Jersey Island and made my trip there possible. **Lieutenant Stan Allen**, the XO of PT- 508, Mairsey Doats, in Lt. Commander Sherertz's Squadron 34, provided much useful background information on Squadron 34 and co-authored the Squadron 34 History. He was on board the PT boat that was intercepted by the British destroyer, HMS Saumarez, when D-Day was postponed due to weather. I had the honor of meeting him at the last two PT boat reunions. **Torpedoman John M. Easterly**, torpedoman on PT- 503, What Next, participated in the rescue attempt of PT- 509. I spoke with him on the telephone and he provided corroboration of various aspects of the rescue attempt, the AWOL chapter, and other Squadron historical facts. **Captain Earl Fox**, XO on Jack Sherertz's first PT boat, PT- 22, the Flying Deuces provided invaluable background about my dad's early time with RON 1. I had the honor of meeting him and his son at the 2009 PT boat reunion in

Arlington, VA. **Quartermaster 2nd Class Howard L. Wollman**, served on PT- 110 in RON 8, John F. Kennedy's squadron. He was awarded a Bronze Star and a Purple Heart for saving one of his boat mates from drowning. I had the honor of meeting him in Murrells Inlet, South Carolina. He corroborated information provided by my father and others about Kennedy. **Fred Hahn, MM2c** served on PT- 156 in RON 9. He had a good friend in RON 2, the squadron that rescued MacArthur, and he helped provide general background information about operational aspects of PT boats during World War II. I had the honor of   meeting him at his house in Winston-Salem NC. I am extremely appreciative of all the help the PT boaters   provided and thankful for their service.

Other assistance with military matters. Alex Johnson is a PT boat historian who provided numerous helpful suggestions about PT boats. I was able to talk with him extensively at the 2009 and 2010 PT boat reunions. Most importantly, he built a Varney model PT boat purchased by my dad during World War II into a replica of Dad's first PT boat, PT- 22, the Flying Deuces. In addition, he read one of the early drafts of my dad's World War II story and made many helpful suggestions. I will be forever grateful for the respect he showed my father and all PT boaters related to their service in World War II. **Marine Sergeant Joe O'Flaherty** is a friend, colleague, and Vietnam veteran. Joe was gracious enough to listen to many questions about the impact of various military events on servicemen and to give me constructive feedback. I am extremely appreciative of all the help he provided and thankful for his service.

Captain Christopher Ohl helped me understand Navy jargon and rank.

John Ovenden is the director, writer, editor and cinematographer of the documentary film entitled, 'PT- 509, The Last Patrol'. The information he provided was invaluable toward giving me a perspective related to where the Jersey Island sea battles took place and the circumstances defining them. I was in awe of the degree to which he had embraced my father's squadron and their men. Thank you, John, for your respect and reverence toward the men, for your assistance with my efforts, and to your family for their hospitality during our visit.

Alyce Newberry Guthrie is one of the driving forces behind PT Boats, Inc. She is the daughter of "Boats" Newberry, the founder of PT Boats, Inc. She provided answers to many questions that came up during the research for this book. I am extremely grateful to all the help she provided and all the respect she showed all of the PT boaters.

University of Iowa 2011 Advanced Novel Writers' Workshop. In July of 2011 I attended a one-week Advanced Novel Writers' Workshop at the University of Iowa run by Susan Taylor Chehak. She and the workshop participants (Nan Brucker, Charlie Drees, Nancy Fraser, Beth Havey, Ryan Kennerly, Michael McLindon, Val Shay, Julia Smillie, David Stevenson, Patrick Trotti, Michael Van Natta) read the chapter in this book on the Pearl Harbor attack and made numerous suggestions to improve it and my overall writing style. Thank you all for your critical suggestions and your encouragement to keep going.

Editorial contributions. William Greenleaf worked hard as my initial editor to make the story better, and I am most appreciative of his efforts. **Loreen Herwaldt**, a long time friend, colleague and co-member of a small writer's group, the Oliver Wendell Holmes Society, read the entire book and made many helpful suggestions. I am most appreciative of her efforts, her encouragement through out the process, and her pushing me to go to the University of Iowa's Advanced Writer's Workshop. **Chuck Diggins**, a long time friend, also read the book and made helpful suggestions. I would additionally like to thank **Kathleen Broome Williams** who read the book quite critically and made many important suggestions. I would like to thank **Adam Nettina** for a number of structural suggestions that strengthened the story. And finally, the last thank you goes to **David Ingram** who assembled the book into its final form.

Miscellaneous. Christopher Nobo spoke with his uncle, who is from Cuba, to find out what the most popular dances were in Cuba during World War II. **Luis Barroso** spoke with his family to help determine the correct spelling of Luis Lombano. **Tatyana Gluzberg** is a professional colleague of mine who speaks Russian as her first language and helped me rewrite the lines spoken by the Russian officers. **Richard Weinberg** helped refine the story concept.

2010 North Carolina Charlotte Flight of Honor, I would like to thank all of those connected with the Rotary Club who made my dad's flight possible. In particular I would like to thank **Staff Sergeant Curtis Winfred 'Wimp' Wise and Aviation Metalsmith 2nd Class Raymond Van**

Havere for their World War II service and the help they gave my dad in the time I worked with them during the Flight of Honor.

To contact the author, send an email to the following address: hjsherertz@gmail.com.

Part I

Jack Sherertz

Sixty-Fifth D-Day Memorial Celebration

Figure 1. Lt. Commander Jack Sherertz (right), USNR, Marine Major
Carl Sherertz (left), 65th D-Day Memorial Celebration.
R. Sherertz collection.

June 6, 2009, Bedford, Virginia

My father, Lt. Cmdr. Jack Sherertz, USNR; his first
cousin, Marine Major Carl Sherertz; my dad's nephew,
Larry Sherertz; and I arrived for the Sixty-fifth D-Day
Memorial celebration in Bedford, Virginia, at 11:00 a.m.,
one hour before it opened. Carl made all the arrangements,
including obtaining a parking sticker that allowed us to
park across the road from the entrance to the Memorial. My

dad was eligible for this parking space because he participated in the D-Day invasion—something I previously knew nothing about. As I slowed my truck to a halt, a slight breeze caused all the US flags in the distance to flutter against a cloudless sky. I looked over at my father, sitting in the passenger seat. He didn't move or say anything.

Finally I got out and walked around to his side of the truck. I opened the door and looked in at him. "Are you ready, Dad?"

The usual twinkle in his eyes had vanished, replaced by a distant look. He showed no signs of being ready.

I was in no hurry. It had taken 54 years after the war for him to ask a family member to go to a World War II memorial celebration, so a few more minutes was nothing.

Finally, he looked at me, took a deep breath, and nodded. I helped him slide out of the truck seat and step down. His thin, wavy, brilliant white hair and deeply furrowed facial wrinkles showed every bit the ninety-one years he carried with him. He wore a white shirt, a navy blue suit, and spit-polished black dress shoes, fitting for a Navy man. Over his left breast pocket, his second wife, Lillian, had sewn his World War II ribbons and medals. From behind my back, I brought out a surprise gift, a navy blue baseball cap, and handed it to him. It said *WWII* and *D -Day* on the front and *Lt. Commander H. J. Sherertz* on the back. He examined it, front and back, and carefully put it on. A big smile followed, accompanied by a single tear running down his face.

He turned toward the Memorial entrance, causing the sun to gleam off his Bronze Star and British Distinguished Service Cross, bracketing his Purple Heart. He walked slowly, stooped over, looking perhaps nine inches shorter than the five-feet-ten-inch height he sported in his early years. We crossed the road that separated the parking lot from the Memorial. Stepping up onto the sidewalk, he stumbled slightly, but quickly steadied himself. In front of us was a tent with a sign that said *Check In Here*. Carl and Larry Sherertz walked ahead to present our tickets, so we could proceed inside.

Before my father and I could catch up, some people saw us and came over. A young boy approached my dad and reached out to shake his hand.

"Thank you for your service," he said, looking up into my dad's eyes with a wide-eyed, unwavering stare.

My father shook his hand and started to say something, but stopped, unable to speak. Tears poured down his face. The boy's parents similarly thanked him and walked away.

We began to walk toward the entrance and got no more than five feet when another group arrived, thanked him, and asked for an autograph.

Then two military men in dress uniform approached us. When they stopped, my dad straightened up as much as he could and pulled his shoulders back, just for a moment, before he had to give in to age and resume his stooped posture. The two officers wanted to have their picture taken with him. He obliged, and the tears continued to flow.

As we neared the monument, a reporter with his cameraman noticed him and came over. He talked to him and tried to answer his questions.

After a seemingly endless stream of people saying thank you, I decided he needed a break, so we sat down in the front row of the chairs arranged for the celebration. The chairs faced the simulated D-Day Normandy Beach landing—a pool of water, steel barricades, and landing craft. Surrounding this D-Day simulation, a group of Marine honor guards in dress uniforms, complete with rifles and pistols, stood at parade rest. They radiated an attitude that said, "No one will hurt you on my watch."

A man with a Vietnam veteran's cap brought Dad a bottle of water. "Do you need anything else?" he asked.

My dad shook his head no.

"Where did you serve?" asked the veteran.

My dad was unable to answer, and more tears followed.

I left my dad with Larry and Carl Sherertz so I could take a few pictures on my own. Finished, I turned to look at him. He stared straight ahead, lost in thought. After a while, he stood up and walked over to the wall surrounding the Normandy Beach simulation. He looked around slowly and then stared into the distance, likely reliving some scene from June 6, 1944. When he turned around, I took his picture. He had a haunted, almost chilling look. He didn't say a word, only stared. Finally he walked back and sat down again.

Figure 2. Lt. Cmdr. Jack Sherertz in front of simulated Normandy Beach landing. R. Sherertz collection.

The rest of the D-Day celebration consisted mostly of politicians thanking the vets for their World War II service. I really can't remember anything they said.

~~~

Soon we were back in the truck returning to Charlotte, North Carolina.

"Dad, how are you doing?"

"Really tired. I must have shaken every hand in the place."

"Some of them, more than once," I said. "Dad, do you think you'll ever be able to talk about World War II?"

"What do you want to know?"

To say I was surprised would be a massive understatement. Not sure that he really meant it, but afraid to miss the opportunity, I asked the question at the top of my list.

"What can you remember about the Pearl Harbor attack?"

He was quiet for a few minutes, staring off into the distance. Just as I started thinking he wasn't going to answer, he began. "I was on leave the night before the Japanese attacked and returned the next morning . . ."

He rambled on for a while, pouring out a stream of seemingly unrelated facts. I didn't interrupt because he clearly *needed* to talk. Then he said something that really hit me hard.

"Bob, my ship, the USS *Nevada*, and the USS *Arizona* were sister ships. They used to tie up next to each other. Their men came to our ship. Our men went to their ship. I had many friends on the *Arizona*. When the *Arizona* blew up, most of my friends died."

I was speechless. Nearly 1,200 men were killed on the *Arizona*. I tried to wrap my head around what I'd just heard. "Is that why you don't talk about the war?" I eventually asked.

He stared into the distance. Finally, he replied. "Do you remember those boxes I asked you to take home and store when I moved in with Lillian? Four of them have things I

8

saved from World War II. I want you to look through them. Then we'll talk."

World War II Artifacts

Figure 3. World War II artifacts from Lt. Cmdr. Jack Sherertz's collection. Clockwise from top left: Squadron history of RON 34; *Victory at Midway,* by Griffith Baily Coale; *Motor Torpedo Boat Manual, Operating Manual for Packard Marine Engine*; marlinspike or fit; Marine knife from Midway with plastic sheath; practice bombs from Midway; Marine gaiter from Midway; Motor Torpedo Boat Squadron 34 stamp; *Seamanship Cards: Rules of the Road*; cardboard blinker to practice Morse code—all sitting on top of his two World War II duffels. R. Sherertz collection.

Saturday after the D-Day Memorial trip, I woke up at dawn, so excited I could hardly stand it. I wolfed down my breakfast, barely tasting it, and hurried out to the garage to look for Dad's World War II boxes. It didn't take long.

I took the boxes inside and placed them on the island in

the center of the kitchen. I stepped back, hesitating. These four boxes likely contained a chapter of my father's life, unknown and undisturbed for sixty-five years. It seemed almost sacrilegious to open them, as if I was disturbing the dead.

Finally, with a great sense of reverence, I slowly drew one of the fragile old containers toward me and pulled gently on the lid. It seemed stuck, so I pulled a little harder. It suddenly came loose, flexing the lid and shooting a half-century cloud of dust into the air. As if I needed anything else to make this moment feel surreal! While the dust slowly settled, I looked at the box's contents.

A slightly yellowed set of cotton scrubs, like those you might see on a hospital orderly, greeted me. I set them gently aside. Next came two large, faded, heavy-duty duffel bags labeled *H. J. Sherertz, Lieutenant, USNR*. Both had heavy brass grommets that at some time must have been threaded by ropes. One of the grommets on the larger duffel bag had a five-inch-long padlock going through it, but no key. I cautiously lifted the bags out and opened them. They were empty, much to my disappointment, but the air inside smelled at least half a century old, amplifying the sense of history.

Below the duffel bags was a rectangular Miller and Rhodes clothing box tied with a string, and a plastic zippered bag labeled *American Society of Travel Agents 1964*. I opened the box first. It was full of newspaper clippings. I picked them up very carefully, one at a time, afraid they might disintegrate at any moment. Most had something to do with my dad's family and friends or community activities that took place during World War II in his hometown of Roanoke, Virginia. It looked as though they were being saved for my dad to read when he came home from the war. I opened the zippered bag and found more of same. After spending another hour going through these clippings, I was no wiser about my dad's war activities.

Books and magazines filled the bottom of the box. The books were all about World War II, but written much later. The magazines were either about World War II or the assassination of John F. Kennedy. I knew John F. Kennedy had been involved with PT boats, but not why my dad included the issues about Kennedy's assassination along with his own World War II stuff. I carefully put everything back in the box and moved on to the second box.

This one was smaller than the first, light brown with faded print on the top that read, *To: Mr. Frank J. Sherertz, Roanoke, Virginia. From: The Riverside Tailoring Company.*

Could it be my dad's uniform? I wondered, with rising excitement.

I opened it carefully and was immediately disappointed. All I could see were more newspaper clippings. Looking more closely, I changed my mind. These clippings were about Dad's war activities. *Japan Attacks Pearl Harbor,* said an extra edition of the Roanoke World-News from December 7, 1941. Just below was a picture of my dad in uniform, the US flag behind him, and someone pinning a medal on his chest. The caption read, *Receives Bronze Star Medal*. Another article, dated February 26, 1941, talked about my dad being commissioned as an ensign. Underneath were clippings with titles that read, *PT Boats Fight with Nazi Minesweepers, Lt. Jack Sherertz Awarded British Distinguished Service Cross by King George,* and *Roanoker on Torpedo Boat in Forefront of Invasion*. I found the clipping at the very bottom most interesting. From the *Charlotte Observer*, it said, *Pearl Harbor Survivor Can Talk About It Now*. I wondered why I hadn't heard anything about this article until I saw the date — December 7, 1999 — one month before my mother died of cancer.

I picked up the third box, labeled *White Bound Paper*, and opened it. The first thing I saw was a pair of lamps made out of two small, stainless steel bombs. Next I found a narrow box about a foot long, which contained a US Marine Corps knife inside a plastic scabbard. Below that was a pair of Marine gaiters labeled *June 6, 1942*. As I kept going, I found a US Navy Morse code training aid; seamanship cards with profiles of different ships; a rubber stamp that, when held up to a mirror, read, *Motor Torpedo Boat Squadron 34*; a piece of wood with a nameplate that read, *Lt. Commander H. J. Sherertz, Supply Officer*; a tapered piece of wood about nine inches long with grooves on the side; and a box with a picture of a PT boat on the outside. PT, I read, stood for Patrol Torpedo. Now I started to get excited because I had heard that my dad had served on PT boats. I opened this box up, and inside were all the pieces

14

necessary to build a model PT boat. Very cool!

Below that were more books. I started to put these aside, thinking they were like the others I had found, until I read the first title: *Restricted Operating Manual, Packard Marine Engine, Model 4M 2500*. It was a manual on how to take care of a PT boat engine. The next said, *Motor Torpedo Boat Manual, US Navy, February 1943*, and had my dad's name on it. Another book was titled *They Were Expendable*, by W. L. White. Inside the book cover were these words: *To Lt. j.g. H. J. Sherertz, MTB Squadron Number One, somewhere in the Pacific, Christmas 1942, with love from Mother and Dad.* The blurb inside said it was about the rescue of Douglas MacArthur. The next book was titled *Victory at Midway*, by Lt. Commander Griffith Baily Coale, USNR, 1944. The author was an artist commissioned by the Navy to document World War II with paintings. I started to flip through the book and look at the pictures until I glanced over at the last book in the box. Its title caught my eye. It said, *At Close Quarters: PT Boats in the United States Navy*, by Robert J. Bulkley Jr, 1962.

Setting the other book aside, I turned to the table of contents. It listed one battle after another, all meaningless to me, so I went to the index. On page 568 was the entry *Sherertz, Lt. Herbert J., 350, 358–360, 449, 477, and 507.* I immediately turned to page 350 where there was a paragraph about Lt. Sherertz leading Squadron 34 PT boats to escort minesweepers for the D-Day assault. On page 358 it talked about Lt. Sherertz becoming the squadron

Commander and went on to describe a battle with German minesweepers. On page 449 it stated that Lt. Herbert J. Sherertz was the Commander of Squadron One from May through October 1943. On page 507 he was listed as having won a Bronze Star.

My heart pounded in my chest, and I couldn't sit still. I took the book to my wife and showed her all the entries. Her eyes told me she thought I had lost my mind.

Undeterred, I returned to the box. Taking a deep breath, I checked what was left and found a diagram of the USS *Nevada*. Using a magnifying glass, I could see that all parts of the ship were labeled. Underneath that was a similar diagram of a PT boat, schedules from 1940 about training exercises on the USS *Arkansas* and the USS *Illinois*, and a magazine called *Yachts* that had articles and pictures about PT boats during World War II. I spent some time studying the diagrams and looking through the books, but I was overwhelmed, because I had no idea what any of it meant.

Then I noticed some videotapes. I thought the first one pretty cool: *First Film Made on Board PTs and USS Nevada*. I picked up the next video, entitled PT- 509: *The Last Patrol*, and turned it over to read the blurb on the back: *PT Squadron 34* . . . I jumped up out of the chair and was about to run out of the room to go view it when I remembered that we no longer had a VHS tape player. I had given it to our son. Momentarily deflated, I went back to see what else remained. The last videotape said very simply, *Jack Sherertz talking with a Library of Congress intern about World War II*.

Just when I thought I couldn't get any more excited, I looked in the box and saw a letter written by my dad's father, Frank, on October 19, 1963. It was a twenty-page, single-spaced letter that he wrote to my two sisters and me about our dad's World War II activities. In it, he said he had pieced the story together from the few things he had learned from my dad and a trip to the Department of the Navy personnel office in Arlington, Virginia. I skimmed it quickly and found out that my dad was first stationed at Pearl Harbor and was on the dock when the attack started. After his battleship, the USS *Nevada*, had to be scuttled, he was reassigned to PT boats and then went to Midway in time for the Battle of Midway. Finally, he was reassigned to another PT boat squadron and participated in the D-Day landing. All three of those battles! Surviving any one of them would have been amazing. Now I was totally hooked.

I was about to reread the letter when something else in the box caught my eye. Staring up at me was a brown folder titled *US Navy Officer Service Record*. At the top it said, *Sherertz, Herbert J.* I put the letter down and reached for the folder. I set it on top of the table, took a deep breath, and opened it. I felt like I just found buried treasure. It was Dad's original World War II service record and contained all of his orders for each assignment—Pearl Harbor; Midway; PT Squadron 34; Warwick, Rhode Island, twice… There was also a copy of an invoice signed by my dad that listed all the things in his USS *Nevada* officers' quarters destroyed by Japanese bombs. Two hours later, I had almost memorized the entire contents. Somewhere in the middle of my reading, my wife brought me a sandwich; I don't remember if I ate it.

I was trying to decide what to look at next when I remembered that I hadn't reached the bottom of the third box. Looking down, I saw a spiral-bound book. I turned it over and couldn't believe my eyes. It said, *History of PT Squadron 34*, by Russell Schuster, Stan Allen, and Sheldon Bosley with a foreword by Jack Sherertz. Dad never told me about this book or about his forward. I quickly read the book, started to put it down, and then read it again.

After that, I looked in the box again and discovered an old newspaper. By this time I had looked at so many newspaper clippings I was almost numb. When I picked it up, I realized that it was only a wrapper for what was underneath—two photo albums. Inside were pictures of my dad and his men—from Midway to D-Day. I also found negatives.

More hours went by.

One box left. It was different from the others—small, but more carefully preserved, with tape over all the joints. I picked it up, sending its contents sliding. Inside was a single, flat, gray metal box. *It must contain something valuable*, I thought. The jammed latch opened after a few minutes of jimmying. Inside were small containers resembling jewelry boxes. I opened them. They were empty but the labels told me they had contained Dad's Bronze Star, his British Distinguished Service Cross, and his Purple

Heart. Another small box contained two patches (still in a cellophane package) with stripes indicating rank, a Navy insignia with an elastic band, double gold leaf pins, miscellaneous buttons, and finally there was a box containing all Dad's ribbons.

Only one of the small boxes remained. Unlike the jewelry-box appearance of the medal boxes, this one was very simple and made out of cardboard, so my expectations were low. I opened it and gasped — it contained Dad's dog tag. I choked up.

Saturday night I barely slept, and I spent almost all of Sunday reexamining the most interesting finds. Finally, Sunday evening, as exhaustion was beginning to close in, I gave my dad a call.

"Dad, I went through the boxes. I have so many questions, I don't know where to start."

"Why don't you come down next weekend?"

"I was hoping you would say that."

~~~

I arrived Friday night late and went to bed before I got wound up and couldn't sleep. Saturday morning we both got up early and had breakfast. Afterward I got the box of things that I had brought with me, and we sat down together at the table.

"Are you ready?" I asked.

Dad nodded.

I put items from the box on the table and my list of questions in front of him.

He looked at the list and thought for a moment. Then, very deliberately, he began speaking. "Why don't I just start at the beginning?"

Part II

The Navy

Signing Up

Jack sat with his father on the side porch one hot and muggy summer evening, their shirts soaked with sweat. The iced tea they were drinking served double duty, both satisfying their thirst and keeping them cool as they held the glasses to their faces. Jack fidgeted in his rocking chair, avoiding eye contact with his father.

"Jack, what did you bring me out here to talk about?" his father asked. "You've seemed quite preoccupied since graduating from Georgia Tech."

Jack took a deep breath and gathered his thoughts. "I've been thinking a lot about the war. Hitler controls most of Europe, and now he's attacking Great Britain. I think it's just a matter of time before Great Britain falls. The United States must get involved!" He paused, looking at his father.

"What makes you think that?" said his father.

"I read in the paper two days ago that Britain has lost one battleship, two aircraft carriers, and three cruisers since the war started. And four days before that, the paper said that Roosevelt had signed a bill to create a two ocean Navy. The bill approved spending a billion dollars. It's going to happen right now, and I want to be part of it. I want to join the Navy."

His father nodded knowingly, a hint of a smile tugging at the corners of his mouth. "What did you have in mind, son?"

"I want to go to medical school. I'd like to start this fall, graduate as fast as I can, and then join the Navy, as a physician. Doc Smith has made inquiries and says I have a great chance of getting accepted by UVA. I've even heard that the Navy might pay my way. When the United States gets into the war, I can go where they need me."

His Father sat silently for several long minutes as Jack sipped his tea, anxiously awaiting his father's reply.

"Jack, I support your choice of going in the Navy, but I'm afraid your timing is bad for medical school. All of my friends who have military contacts in Washington think we're only months away from entering the war. If you've just started medical school and the war starts, the military can pull you out and put you where they want you, very likely in a field position. My advice would be to sign up for officers' training in the Navy, and then you'll have some choice where you go. You can go to medical school after the war."

"Your friends are pretty sure about this?"

His father nodded.

"I've been afraid that might be the case." Jack paused for a few minutes. He thought he had considered every possibility, but he couldn't think of a way around this. "I guess I'll have to wait until the war is over," he said, his voice heavily laden with emotion and disappointment.

"But Dad, the Naval Academy takes four years, so won't that take too long? Seems like I would miss the war. Now what am I going to do?" He said to no one in particular.

"Until the end of June, Annapolis was the only option, but FDR just signed off on the V-7 program," said his father. "It will train officers in three months, after a one month qualifying cruise. The Navy wants to triple the number of officers in the Navy."

"I don't know, Dad. That sounds a little screwy to me. I'll have to think about it. Thanks for listening." He got up and walked slowly back into the house.

~~~

Jack and his longtime friend J. P. Saul Jr. walked along Brandon Avenue, heading east toward the Raleigh Court area of Roanoke. It was a beautiful summer morning with a cloudless blue sky. A gentle breeze and shade from the many trees had so far held off the heat of the day. They reached Grandin Avenue and hung a left, heading toward the Grandin Theatre, where they'd seen many a movie. They walked for a distance without talking, Jack oblivious to the fact that J. P. was staring at him.

"So Jack, your dad didn't go for medical school, did he?" said J. P. "I could tell as soon as I saw you this morning. Now what are you going to do?"

"J. P., after careful deliberation, I have a new plan."

25

Suddenly Jack ran out to the middle of the street in front of the Grandin Theatre and jumped on the back of a streetcar just as it began to move, heading toward downtown. J. P. ran to catch up and jumped on next to Jack.

Jack loved riding the streetcar. It had a rhythm to it that you could see and feel and hear—people coming and going, the crackling of electricity from the contacts above, the clacking of the wheels going over the rails, the conductor's bell. But the best part was the view.

"Jack, there's the car club on their Saturday outing, crossing behind us," said J. P. "Oh baby, I'll take that fire-engine-red 1939 Lincoln Zephyr coupe."

"Very nice, J. P.," said Jack. "But I'll take that jet black 1939 Graham convertible with the white sidewalls. Much classier."

"That '36 yellow Cord 810 is pretty slick. I could see me and a date out cruising in that."

"Gotcha beat, J. P.," Jack said. "I'm gonna take that '35 Auburn 851 Speedster. Hubba-hubba." He made the outline of a curvaceous lady in the air.

"That is definitely the winner," said J. P. "I got shotgun."

They went over a hill, and that was the end of the car show, so they switched to another game: looking for the funniest sign.

J. P. pointed out a sign in a vacant lot, advertising tequila. "Sparkle Donkey. You've got to be kidding!"

They both laughed.

Jack pointed out a sign on the side of a building for Schaefer Bock Beer, because a Dr. Seuss character was the spokesperson. *Aha*, said a sheep-like creature.

J. P. didn't react much, so they kept looking.

Then J. P. pointed out a large billboard showing a penguin on skis coming down a snow-covered slope with a cigarette in its mouth. *Smokers discover it's a swell idea.* Again, they both laughed.

"What about that one, J. P.? *Get Crunked! It strengthens nerves and ensures restorative sleep.*"

"Get Crunked, Jack."

"Get Crunked, yourself, J. P."

They both howled with laughter. Suddenly they were downtown, and signs were scarce.

J. P. couldn't take it any longer. "So what's your plan, Jack?"

"Ask and you shall receive," said Jack, as he jumped off the streetcar. He turned and made a sweeping gesture with his hand from the direction of the streetcar and finished by pointing directly at a building in front of them.

J. P. jumped off right behind him and looked up. They had arrived in front of the Navy recruiters' office, and Jack was pointing toward a line of at least ten men at the door.

"J. P., I've decided to join the Navy and become an officer. Are you interested?"

J. P. hesitated for a moment and then said, "Sure, why not? If I don't volunteer, I'll probably get drafted."

Jack looked at J. P., who had made the decision to join the Navy in a few seconds, while he had been thinking about it for months. *To each his own*, he thought. Though trying hard to appear in control, he churned inside. "J. P., my dad said there would be a lot of volunteers here today after that article about the V-7 program in the paper over the weekend, but I didn't believe him."

The men in front of them were all about their age and sounded nervous as they talked quietly among themselves. Jack felt the same way.

"Where do you think they're from?" asked J. P.

"Let's find out," said Jack. He and J. P. walked over to the back of the line, and Jack tapped the fellow immediately in front of them on the shoulder. "Where you from, sailor?"

"Me and my two friends are from Roanoke. I'm Bill Hall, but people call me Ed. These are my friends, Frank Long and Fred Taylor."

"Where are the rest of you from?" Jack asked, speaking loudly enough for the whole group to hear.

"Troutville," said one.

"Blacksburg," said another.

"Buchanan." "Pulaski." "Galax." "Abingdon." "Lynchburg." "Cave Spring." "Martinsville."

"My name's Jack Sherertz, and this is J. P. We're both

from Roanoke. I hate to break it to you, but you guys are in the wrong line."

For a moment, they all looked concerned.

"This is the line for officer candidates. Some of us have already been in training for this, and I can see that none of you are ready. You hicks should be in the line for enlisted men at the back of the building."

"What do you mean, hicks?" said one of them, who edged forward menacingly.

"What training?" another asked.

"Navy ROTC during my fourth year of college," said Jack. He looked at them with a straight face for a moment and then started laughing. "Just kidding. Welcome to Roanoke."

That seemed to break the tension, and conversations got louder and more energetic.

"I thought for a second he was talking about the line for the colored," said one of the young men to his friend. "Then we really would have had a problem."

"What's the problem?" asked his friend. "Don't you want to be peeling potatoes with the pickininis?" They both laughed, with most of the others joining in.

"Pickininis," Jack said to himself. *Why would they want to fight in the war after the way we've treated them?*

In a few minutes, a young man in a navy blue uniform and a white hat tipped to one side came out and handed them all a sheet of paper. "I don't want any of you waiting in line unnecessarily," he said. "Read the information about the V-7 program and be sure you're qualified and ready to do what is necessary."

"OK, this doesn't seem to bad," said JP. "Birth certificate, unmarried, pass a physical exam, two years university, two letters of recommendation; we can do this, Jack. But what's this part about being in the reserves? Will we be active duty or what?"

"Yes, you'll be active duty," said the Navy man over his shoulder, as he walked back inside.

"No pay," said one of the non-Roanokers. "What's the deal with that?"

"But I hear the cruise is to the Caribbean," said another. "That beats the heck out of baling hay all summer."

"J. P., this is going to be exciting," said Jack. *Exciting in a life-or-death kind of way, he thought. I wonder how many in this line will survive the war. I wonder if I'll survive the war.*

"What do you want to do after we get through the V-7 program, Jack?" asked J. P.

"I want to end up on a battleship and fight the Germans in the North Sea," said Jack. "One of my Navy ROTC instructors said that the Germans are building battleships with 15 inch guns that can fire 20 miles. I would like to be part of a battle like that."

Those around them looked suitably impressed, hearing about these massive battleships. Then someone asked,

"Why do they call ships she?"

"Who would you be more afraid of, a ticked-off woman or a ticked-off man?" Jack replied.

They all laughed.

Just then the same Navy man who had come out earlier opened the door and said, "If you're interested in the Navy, come on in and sign up."

~~~

"Jack, that was easy. I wonder how long we'll have to wait for our cruise."

"Based on some things I heard my dad and his friends say last night, I think we may be waiting awhile. The program's barely off the drawing board."

Figure 4. Jack Sherertz (left) and friend in Georgia Tech
Navy ROTC uniforms. Jack Sherertz collection.

Figure 5. Jack Sherertz in Georgia Tech ROTC drills. Circa 1936 or 1937.
Jack Sherertz collection.

USS Arkansas Training Cruise

Figure 6. USS Arkansas. Wikipedia.

September 30, 1940

Jack and J. P. stood at the front of about 1000 apprentice seamen in the Brooklyn Navy Yard adjacent to the USS *Illinois*. Some looked like they just got off a hay wagon in their jeans and work shirts. Others had on a coat and tie. All looked less than thirty. As Sherertz and J.P. listened to the conversations going on around them it became clear that this crowd of young men came from all over the

United States.

On a bulletin board set up next to the ramp leading up to the USS Illinois were two postings. The first said *Apprentice Seamen Cruise Itinerary, and the second said Pointers for the Guidance of the V-7 Reservists.*

A loud voice near them said, "A cruise to the Caribbean, with stopovers in Cuba and Panama. This is going to be a piece of cake." Nervous laughter followed.

> Apprentice Seamen Cruise Itinerary:
>
> Depart NY Tuesday 10/1/40, Arrive Guantanamo Saturday 10/5/40
>
> Depart Guantanamo Wednesday 10/9/40, Arrive Colon, C. Z. Friday 10/11/40
>
> Depart Colon, C. Z. Monday 10/14/40, Arrive Norfolk Saturday 10/19/40
>
> Depart Norfolk Friday 10/25/40, Arrive NYC Saturday 10/26/40

"J. P., I think this is going to be a lot tougher than it looks on paper," said Jack. "I wrote to one of my Navy ROTC officers at Tech, and he wrote back that all regular Navy officers hate the V-7 program. They're going to be looking for any reason to get rid of us." *I wonder if I have what it takes,* he thought.

POINTERS FOR THE GUIDANCE OF THE V-7 RESERVISTS

Always be in the uniform of the day.

Wear your uniform properly, hat square on your head.

Do not leave clothing adrift.

Be prompt to stand by your bedding at "Hammocks," particularly when it is "piped down" after being aired.

Bedding must be stowed or lashed in your hammock.

Be sure your bedding is properly secured to the lifelines or rails when it is aired.

Do not lean on lifelines at any time.

Keep your locker properly secured at all times.

Anyone finding or losing any articles shall report the fact to the AIDE in the reserve office.

The Luck Bay (bag kept by the Chief Master-at-Arms for stowing of loose articles found about the ship) will be open certain hours each day.

Clear the deck of mess gear immediately after you have eaten your meal.

Do not throw cigarette butts or other trash on deck, over the side, or in the urinals and scuttlebutts.

Compartment cleaners must air their deck swabs topside daily.

DO NOT WASTE FRESH WATER. It is made aboard ship and must be used with the greatest discretion.

Do not use profane, lascivious, or obscene language.

Do not whistle on board ship. It is unseamanlike.

Observe all safety precautions.

At emergency drills move quietly on the double. Go forward on the starboard side. Go aft on the port side.

Make it a habit to read the plan of the day posted on your division bulletin board.

Obey cheerfully and promptly all orders of those placed in authority over you regardless of their rank or rate.

If you are in doubt about anything, do not hesitate to ask questions.

Keep in mind at all times that you are under constant observation by the officers and petty officers of this ship. You have presented yourself for a commission, and it is up to you to prove by your bearing and attitude that you are worthy of that commission.

Never let yourself be a party to the spreading of gossip and false rumors.

Never let yourself harbor a grudge for having been chastised by an officer or petty officer.

Your captain, the officers, and petty officers over you will exert every effort to make your stay aboard this vessel as pleasant and as instructive as possible.

Just as the crowd of young men started to get restless and mill around, they all heard a distinct sound coming from a bosun's pipe in the mouth of a tough-looking seaman at the top of the USS *Illinois* ramp. When everyone got quiet, he dropped the pipe out of his mouth, letting it swing from a lanyard around his neck. He came down the ramp with a swagger that caused everyone to take several steps back. Jack could see the crossed anchors on the right arm of his white uniform.

He looked everyone over and then spoke with a loud New York accent. "I am the chief bosun's mate on the USS *Arkansas*. For the next month, your asses are mine! The USS *Arkansas* is *my* ship. If you do not do everything I tell you, exactly like I tell you, you will not be graduating. Half of you swabbies won't make it to Midshipmen's School. You address me as sir. That sound you just heard means 'everyone assemble on deck.' Is that clear?"

"Yes, sir," the recruits mumbled.

"I said, is that clear?" the bosun's mate yelled.

"Yes, sir!" they all answered, louder this time.

"Now yous guys, if you'll all follow me, we'll walk across the USS *Illinois* to the launches waiting on the other side and get all of you maggots over to the USS *Arkansas*. You'll get your uniforms and the rest of your issue when you get on board. One more thing: the hammocks are first come, first serve."

Off he went, apprentice seamen trailing behind.

~~~

On the *Arkansas*, the bosun's mate herded the new seamen to the foredeck and then into lines facing the forecastle. Once assembled, he spoke. "The skipper on this ship is Captain John L. Hall Jr. Pray that you won't have to speak to him until you graduate. Today you'll get your sailing orders from the executive officer, Commander P. M. Rhea. Attention!"

The men attempted to stand at attention, but failed miserably. Appearing very apprehensive, they waited.

Commander Rhea walked over and stood in front of them, slowly looking them over. The longer he looked, the more they sweated. When their uneasiness was at its peak, he began speaking.

"Gentlemen, if you are lucky, you will become part of the new Navy. What is the new Navy, you might ask? It is a Navy that thinks you can train an officer in ninety days. They call them ninety-day wonders. Most of us regular Navy men don't believe that's possible. So we're going to be looking for any reason to ship you out. If I catch any of you acting disrespectfully aboard the USS *Arkansas*, I'll throw you overboard myself. She served in World War I and is now the flagship of the training squadron. At this time I'll turn you over to Captain Ruffin, the Commander of our training school, who will start your transition into becoming Navy men." He turned and left without looking back.

Unlike Commander Rhea, the Captain wasted no time in getting started.

"Two days ago four of our destroyers flying British colors arrived in the UK to help the Brits against the Germans," said Captain Ruffin. "That's just the beginning of what we're going to be doing in the North Atlantic. We also believe that the Japanese are going to make a move in the Pacific, so there will be many opportunities for those of you who graduate. Let me elaborate . . ." He droned on interminably, failing to notice that everyone had tuned him out. Finally, he yelled, "At ease. Dismissed."

~~~

The next morning, reveille sounded at 0500, followed by some person coming through the hammocks yelling,

"Calling all idlers."

No one needed to be awakened. Everyone had been up most of the night, seasick due to rough seas. Those not seasick had to listen to the rest, so no one got any sleep. The hammocks rapidly emptied as everyone went into the head, wishing for a 3S – shit, shower and shave – but having time only for a PS – piss and a shave. A line quickly formed behind each sink and mirror, so some tried to shave at the scuttlebutt. That's Navy for water fountain. A few got caught at the fountain by a bosun's mate and were reamed out. Everyone was exhausted, white-faced, or both. If that wasn't bad enough, everyone's arms were sore from receiving inoculations against infections, the night before. A few minutes later, the bosun's mate piped assembly on deck.

Day one meant fire drills. The bosun's mate put each seaman into position and showed them how to use the hoses. Then the apprentice seamen turned on the hoses to spray down the decks, but one hose got away. Acting like a giant snake thrashing about in the air, it soaked everyone on deck. Off to the side, another bosun's mate stood with a clipboard, writing down comments about everything being done.

"Now that you've got my deck wet, you might as well clean it," bellowed the bosun's mate. "You're going to holystone it until it's clean and white. Since you're all dumber than jellyfish, I'll demonstrate how it's done."

He took what looked like a broomstick and stuck it

through a hole in a well-worn brick sitting on the deck surface. Then he took a mop from a bucket full of soapy water and washed a small area. He grabbed a handful of sand from another bucket and threw it out on top of the soapy area. Finally, he pushed the brick around with the broomstick, and there was a loud grinding sound as it moved back and forth.

"If done properly, my teakwood decks will all gleam white. If done improperly, you will stay here until they do. Or I will throw you overboard and let a U-boat or a shark take care of you. Is that clear?"

"Sir, yes, sir."

~~~

After cleaning the deck, each of the eight divisions and their squads were ordered to go read their schedule for the day. J. P. Saul and Jack Sherertz were in Division 6 and saw the following.

October 2, 1940 – Divisions 5 & 6

0545   Call all idlers

0615   Sweep down, lead out, wash deck hose. Wash down with sand and soapy water.

0630   Reservists N. 41 to 80 incl. report to sick bay to leave specimen of urine.

0700   Dry down. Up all hammocks.

0800   Turn-to; ship's work. Two hands from each of the first eight deck divisions report to the boatswain's locker for duty as a cleaning gear breakout party. Catapult no-load shot. Warm up anchor engine. Flight quarters.

0830   Quarters for reservists. Hoist in all boats, rig in all booms, trice up gangways. Catapult one plane.

0915   As soon as boats are hoisted, hoist plane aboard.

0930   All reservists receiving bismuth injections report to sickbay. School and drill call for reservists. Station all special sea details.

1045   TC Hudkins, Priest, Brown

Section 1 Turret Instruction

Sections 2, 3
Broadside Instructions

1315   Same as 1045

1400   Lt.   Commander   Criddle Organization   of   Gunnery Department

Lt. Martineau
Description of Armament

1500 Capt. Ruffin

      Articles for the Government of the Navy

      Second Lt. Collins
      Limitations of Punishment

1600   All reservists to sick bay for inspection of vaccinations.

1930   Movie: *The Gracie Allen Murder Case*, with Warren William, Gracie Allen, Kent Taylor, and Ellen Drew.

After reviewing the schedule, the men took off for either sick bay or to take care of their hammocks.

~~~

At the end of the day, only a few of the five hundred seamen tried to watch the movie, and those fell asleep. Finally they all heard the bosun's pipe followed by, "Pipe down and lights out."

That second night, there wasn't much conversation after lights out, but sleep still didn't come easy to the exhausted, apprentice sailors. All night, they could hear the hammocks squeaking, as men tried to get comfortable with their sunburn, or the sound of feet hitting the floor as sailors made a mad dash to the head because of seasickness.

~~~

The next morning, Jack and J. P. checked out the day's

schedule posted on the Division 6 bulletin board.

"Turrets, broadside firing, boatswain and bugle calls, painting the ship, marlinspikes," Jack noted. "What's a marlinspike?" He looked around for an answer.

No one had one. They all shook their heads.

"I guess we'll find out."

Jack, J. P., and the others first went up in a turret to learn how it worked, inside and out, and how to troubleshoot if it wasn't working. Then they moved on to broadside firing. The classroom session was all about the physics, ballistics, and explosives involved with firing the three-inch, five-inch, and twelve-inch guns of the *Arkansas*. But then came the part they were all excited about: firing the twelve-inch guns. A gunnery sergeant took their division up inside one of the twelve-inch gun turrets. He sequenced the firing, from how the powder and shells came from below deck on conveyor to how to load them, lock them in the chamber, and fire them. Then they actually loaded a shell, rammed in the powder, and closed and locked the breach, so the gun was ready to fire. Next they went below deck to the magazine to see how the powder and shells were stored and transported. Finally, it was time.

"Everyone follow me up to the gunnery observation deck," said the gunnery sergeant. When they were all assembled, he reached for a ship phone that went straight to the turret they had just left. "Cover your ears, gentlemen. Fire when ready . . . "

Boom!

"Holy shit," said one of their division seamen. "The recoil from that shell moved us back a foot."

"If we were giving a full broadside, it could move the boat sideways up to ten feet," said the gunnery sergeant.

And so the day went until just before dinner mess call, when Jack and J. P. were rehashing the day's events.

"Bosun's Mate O'Connor was pretty amazing with that marlinspike," said Jack. "None of us were able to get any of those knots loose, but he did it in seconds. J. P., I'm going out on deck to practice while it's still fresh in my mind. He said no self-respecting seaman would be caught dead without his marlinspike on his person at all times."

"Suit yourself, Jack," said J. P. "I'm going to go put my feet up for a few minutes until chow time. Can you believe how excited Boatswain Kuraskiewicz was about teaching us how to paint the ship?"

"Someone's got to be excited about it, or our boats will turn into rust buckets," said Jack. "But I really liked Boatswain Russell's pipe and bugle calls. I think I could dance to that music." Jack smiled.

J. P. shook his head and walked away.

That night, Jack's level of exhaustion was so high that if anyone got seasick, he didn't hear it. He slept through everything.

~~~

The night before their arrival at Guantanamo Bay, Cuba, Jack and J. P. stood on deck on either side of a hatch that led to stairs going below. J. P. had his hands in his pockets. With his right arm, Jack held on to a lever above his head that was used to seal the hatch when it was closed. In the distance, they could see the USS *New York*, silhouetted against the moon, also inhabited by 1000 apprentice V-7 seamen. At the sound of footsteps coming up the stairs from below, J. P. turned to see who was coming through the hatch.

Surprised to see that he was flanked on either side, the exiting seaman recovered quickly. "My name's Klinknett. I'm writing an article for the *Arklite*. Anyone send you to the engine room for a bucket of steam, or ask you to find a hammock ladder?" he chuckled. "What do you plan on doing in Cuba?"

Suddenly Jack's knees buckled, and he fell toward the deck. Just as his right arm reached maximum extension, he jerked to a stop and hung there, holding on to the lever. "Must have fallen asleep," he said with a big grin. "I think I need to get my blood moving. A little dancing should be just the ticket." With that, he grabbed a mop and danced around the deck.

"What kind of dance is that?" Klinknett yelled after him.

"Mambo," Jack called over his shoulder.

Klinknett watched, transfixed.

After a few minutes, Jack stopped and walked back over

to join them. "Why do you want to write for the *Arklite*?" he asked.

"Actually, I want to write for *Stars and Stripes* once I'm finished with Midshipmen's School, so I can get a job with a newspaper after the war. What are you guys going to do when the war's over?"

"I'm going to medical school," said Jack, with some excitement.

"I don't have a clue," said J. P.

"So who's going to dance with you in Cuba? I'll bet you can't speak Spanish."

Jack and J. P. looked at each other and laughed.

"Back home in Virginia, we call him Lady's Man," said J. P. "He won't have any trouble finding a dance partner. Jack, I've had enough fun for one day. I'll see you in the morning."

J. P. stepped through the portal and disappeared down the stairs.

"Well, I'm still looking for a story," said Klinknett, and he headed off to find someone else to interview.

Jack went back to dancing around the deck. A few minutes later he heard, "Yo, Fred Astaire, where's the party?"

Jack turned toward the voice and saw a small group of apprentice seamen watching him. "Who wants to know?" he asked.

"I heard a rumor we have something in common," said one of the seamen.

"Do you dance?"

"A little, but that's not it."

"What, then?"

"We all like battleships."

Jack stopped dancing and walked toward them.

~~~

Saturday morning they arrived in Guantanamo Bay, Cuba. The men had carefully washed their uniforms the night before and hung them out to dry on the lifelines ringing the deck perimeter. Although slightly wrinkled, the brilliant white clothing looked good on the mass of apprentice seamen. Jack and J. P. stood on the bow, watching the ship maneuver in to the dock. Suddenly they were joined by a group of similarly dressed men.

"J. P., I want you to meet the BB Wonders," said Jack.

"BB Wonders?"

"I met these guys last night," Jack said. "They—uh, we—all want to end up on a battleship."

"...Would you accept another member?" asked J. P., after some hesitation.

"Absolutely," said Jack, and the rest nodded. Jack was beginning to wonder if J. P. really liked battleships. "J. P., this is Walter Drane from Cleveland Heights, Ohio. He went to the University of the South in Sewanee, Tennessee.

He's thinking about being a minister, so everyone calls him Rev."

"Nice to meet you, Rev," said J. P., as he shook his hand.

"Next is Andy Uhrenholdt, from Hayward, Wisconsin, near the Great Lakes. He came from Ely Junior College. This is Tommy Jones from Tallulah, Louisiana, near Vicksburg. He went to LSU and actually thinks they can beat Georgia Tech this year. And last but not least is Johnny Mote from Spiceland, Indiana. Spiceland has eight hundred people, and Johnny knows them all, or so he says. He went to Purdue."

When J. P. finished shaking their hands, he asked, "What's on the agenda for today, Jack?"

"I've got a big surprise."

~~~

After they disembarked, most of the seamen headed toward a line of buses there to take them to town, but Jack and his companions kept walking past the buses to a civilian parking lot. As soon as they got close, a young man dressed all in white came running up to Jack, talking a mile a minute, and gave him a big hug.

"Everybody, meet Luis Lombano, a school chum of mine from Georgia Tech who lives in Havana. He's going to take us to dinner and dancing in Santiago de Cuba. We're going to have a great time."

Luis bowed with a flourish and enthusiastically shook their hands. "It is a pleasure to meet all of Jack's friends."

"I like this guy," said Tommy, as Luis pumped his hand up and down.

They all piled into a big black car with a canvas roof and took off.

Figure 7. Apprentice Seaman Sherertz and buddies next to Luis Lombano's car. Sherertz is 2nd from right in the lower row. Jack Sherertz collection.

~~~

When they arrived in Santiago de Cuba, it was early evening. The sun, low in the sky, cast a golden glow over all the buildings. Many had second-story balconies, bordered by fancy ironwork that cast sharp shadows, especially in the archways. Luis drove them to a restaurant near the harbor, and they were seated outdoors in a courtyard that overlooked the bay. It was a pleasant 80 degrees with a gentle breeze blowing in off the water. Luis spoke briefly to the waiter, who disappeared and came back with a full pitcher and some glasses.

Luis stood up, took the pitcher, and poured everyone a drink. "Welcome to Cuba," he said. "I warn you, these drinks can put you on your ass."

"This is delicious," said Rev. "What's in it, Luis?"

"Rum, club soda, lime juice, and sugar. This place makes one of the best mojitos in Cuba."

Everyone enjoyed the fresh air and being off the ship. While the others drank up, Jack and Luis caught up on the last five months since graduation. None of them noticed it getting dark. They were all having too much fun on their first leave. Outdoor torches had been lit, and the flames moved in a lively fashion, blown by the offshore breeze.

After Luis spoke with the waiter again, he said, "Gentlemen, I took the liberty of ordering some of our local favorites. We'll be having *Moros y Cristianos*, aka black beans and rice; *ropa vieja* with *criollo* sauce, aka shredded beef; *viandas fritas*, aka fried potatoes; and *plantains*, aka fried bananas. Enjoy!"

From behind, they heard a group of women speaking Spanish and giggling. They turned to find them coming in their direction.

Luis stood up and bowed to the women before turning back to Jack and his friends. "When Jack told me he was coming and wanted to go dancing, I decided to bring some of my lady friends. I would like you to meet Lola, Ade, Elena, Carmen, Sophia, and Mariela."

Jack and Luis pulled over a table and helped the women get seated. They each smiled shyly, and it was clear that most of them spoke very little English. Unable to communicate, no one seemed to know what to do next—until the music started. Luis took Elena's hand, and they walked to a central area and began dancing. Jack and Mariela followed. The remaining guys sat and watched.

"Jack, what kind of dance is that?" yelled Mote. "I've never seen anything like it."

"Mambo," Jack yelled back. "Luis taught me."

Each of the other women stood up, walked over, took one of the guys by the hand, and headed to the dancing area. Even those who couldn't really dance seemed to be having a great time. When the band finally took a break, the couples all sat down and had glasses of water.

After they all finished their dinner, Jack stood up and bowed with a flourish, prompting a big smile from Luis and his lady friends. "Luis, it is unfortunate, but we have a curfew. It has been a pleasure as always, but can you take us back to the base? *Senoritas, talvez amanha.*"

The ladies smiled. The guys gave them hugs, paid the bill, and went out to Luis's car to return to Guantanamo Bay.

~~~

Sunday morning, Jack and J. P. were standing in front of the bulletin board looking at the day's schedule when an E7 came by and stuck something new on the board.

"Hot off the press, gentlemen. Yesterday's football scores." He said it loud enough that soon there was a crowd discussing the results.

The BB Wonders huddled together, actively discussing the scores.

"Why doesn't Tech pick on someone their own size, and leave poor Samford alone," J. P. asked Jack.

"You're just upset because UVA barely beat Yale," said Jack.

"A good second week for your Tigers beating NC State, Tommy," said Mote.

A voice from the back of the gathering crowd said, "Go, Navy!" Everyone cheered.

Then a bosun's mate piped chow time.

"Let's go garbage up some bloody powdered eggs and spam. Yum!" said another voice from the back, and everyone took off to eat.

~~~

Friday, October 11, the seamen arrived in Colon, Panama. After departing the ship, they all got on a train bound for Balboa across the Isthmus of Panama. The BB Wonders established themselves at the end of one train car, and Jones pulled out a deck of cards.

"Five-card stud," he said. "Ante up, boys."

"Did you catch a look at those women watching us as we pulled out of the train station?" Uhrenholdt asked. "I think I'm in love."

"You thought you were in love in Santiago, you sentimental sap," said Mote.

"I bet I get a date for Saturday night before the rest of you," said Uhrenholdt.

"How much?" asked Jones.

"I got five bucks," said Uhrenholdt.

"I'll take a piece of that action," said Mote. "My money's on Sherertz."

"Put your money where your mouth is," said Uhrenholdt.

"Why don't we see who still has money after the poker game?" said Jones as he began dealing the cards. "Two bits is the ante."

~~~

Later that evening in the train station Uhrenholdt could be seen paying off those who matched his $5, as everyone but Uhrenholdt smiled.

~~~

On the way back to Norfolk, October 18 started out slow, like every other day, but heated up in the afternoon. The excitement began with a "man overboard" drill, to teach the seamen how not to panic should they ever fall overboard. They formed two lines, the first at the starboard bow, where each seaman in turn had to jump into the water. The second line was located midship at the top of the starboard gangway. At the bottom of the gangway was a launch tasked with rescuing each man overboard. In charge of the launch was a very large seaman nicknamed Frankenstein.

Jack was in the bow line, waiting his turn. The seaman in front of him was nicknamed Rabbit, because he always seemed nervous and kept looking around. Today the nickname seemed appropriate, because Rabbit appeared increasingly anxious as his turn to jump neared.

Concerned, Jack asked, "How's it hanging, Rabbit?"

"Are you kidding?" asked Rabbit, a terrified look on his face. "I'm about to jump twenty-five feet off a perfectly good boat and then use my nonexistent swimming skills, and you want to know how I'm doing?"

Now Jack was really concerned. He went over to the bosun at the head of their line. "Rabbit can't swim," Jack said.

"I guess we'll see in a few minutes, won't we?" said the bosun. "Get back in line, sailor."

"Everything will be fine, Rabbit," said Jack, but inside he didn't feel it. The look on Rabbit's face said it all.

When it was Rabbit's turn, the bosun had to push him overboard, because he wouldn't jump. As everyone turned to watch Rabbit in the water, Jack got out of line and followed him as he floated toward the starboard gangway. Initially Rabbit did okay, but when he approached the launch, the launch didn't move to intercept. Looking closer, Jack realized the men couldn't get the launch untied, and Rabbit floated by, just out of reach. Frankenstein yelled out orders, sounding frantic, and Rabbit was beginning to thrash about in the water. Jack heard Frankenstein yell something about cutting the rope and knew it was going to be a few minutes before the launch reached Rabbit, so he kept following Rabbit as he drifted toward the stern. Less than a minute later, Rabbit floated into the wake of the *Arkansas*.

"He's in trouble," an observer in the stern yelled through his megaphone.

Without hesitation, and before anyone could stop him, Jack stepped over the lifeline and dove into the wake. He came up swimming fast and reached Rabbit in a matter of seconds. "Looks like your ride's a little late," said Sherertz. "What say I practice my lifesaving skills?"

Jack's presence seemed to calm Rabbit, and he stopped thrashing. Jack got behind him and initiated a cross-chest carry. By the time the boat arrived, Rabbit had relaxed and stopped coughing, and they were quickly pulled into the boat. Jack and Rabbit sat down on the seats.

"Thanks, Jack," said Rabbit. "You saved my bacon. Where'ja learn to dive like that?"

"Cliff diving in the Rockydale limestone quarry near Roanoke," said Jack.

"Well, I don't know where that is, but I'm glad you were looking out for me."

"Someone's going to pay for this," Frankenstein said, loud enough for everyone in the boat to hear.

"What's his problem?" asked Rabbit.

"He probably fell out of his hammock this morning and

hasn't recovered," quipped Jack.

"Do you suppose someone pulled that rope tight on purpose?" Rabbit whispered.

Jack shrugged but smiled knowingly.

When they got back to the starboard gangway, a bosun's mate was there to greet them. "Nice work, Seaman Sherertz. Everybody topside, and we'll talk over what just happened."

Frankenstein winced, but he seemed to be gaining control of himself by the time Jack and Rabbit stepped off the boat. Then Jack saw the knot that still remained where the rope had been cut. He reached into his pocket, pulled out his marlinspike, and jammed it into the knot. In a matter of seconds, he had the knot untied. Behind Jack and Rabbit, Frankenstein shook with anger.

The bosun's mate went over the exercise and lauded Jack's quick thinking, emphasizing that Frankenstein's delays could have been a problem if Jack hadn't reacted so quickly. Frankenstein continued to seethe. He might have quieted down except for the next activity: greased-pole boxing.

The game was simple. A five-inch pole was suspended over the deck. Two men were put on the pole facing each other, a boxing glove on their dominant hand, the other hand bare to steady themselves on the pole. Typically the game was played until one man knocked the other off the pole.

Usually it was great fun, but today the first one to volunteer was Frankenstein. Right from the beginning, he acted as if he intended to hurt someone. After the second opponent slammed to the deck, no one else seemed interested in facing him. Frankenstein was at least six feet six inches tall and a mass of muscle from rowing crew at some Ivy League school. Just when it seemed as if the game might be over, "Slugger" White stepped up and challenged him. Slugger was a professional boxer.

Everyone started cheering, but the gunner's mate shook his head no. "Anyone else?" he asked.

"Isn't that the jerk that keeps stealing all the prime hammock spots?" Jack asked J. P.

"You got it."

Jack walked straight over to the gunner's mate running the exercise and talked to him a moment. The gunner's mate nodded and smiled. Jack winked at J. P. and walked to the center of the ring.

As soon as Frankenstein saw Jack, he fairly licked his lips.

Jack turned to Frankenstein. "I would like to propose a different game and see if it meets with your satisfaction," he said. "I like boxing better when your feet are on solid ground and both hands are gloved." Loud cheers erupted.

"You're on," replied Frankenstein, dropping down to the deck with great anticipation. Here was an opportunity to avenge his earlier embarrassment. "Get your butt over here, Seaman, and I'll give you a free pass to sick bay."

Jack was five feet ten inches tall and solid, but eight inches shorter than Frankenstein. Yet Jack didn't seem the least bit worried.

"Ten bucks on Sherertz," said one of the Jack's friends.

"I got ten on Frankenstein," said another seaman. Everyone quickly took a side.

Meanwhile, both boxers were gloved. All the seamen formed a ring around them, crowding in as close as possible. Everyone sensed blood.

The gunner's mate placed both boxers in the center of the ring with their gloves touching. "Ready, box."

Frankenstein immediately stepped forward and threw a massive left designed to take Jack's head off. To his surprise, he hit only air. Meanwhile, Jack had sidestepped to his right, and in the same motion threw a short, powerful counterpunch with his left hand that caught Frankenstein in the ribs under his left arm.

"Crunch," could be heard by all.

"Oooof," grunted Frankenstein. He recovered, and threw a roundhouse punch with his right arm.

Jack had already moved left, and let fly another punch with his right hand that caught Frankenstein in the face, staggering him back and bloodying his nose.

"He bleeds," yelled a seaman, and the crowd moved in closer.

Although Frankenstein was significantly diminished, his anger fueled a second wind. He began throwing punches wildly, chasing Jack around the ring. "Stand still, you coward," he yelled.

Finally, with Frankenstein obviously tiring, Jack let him come straight at him. Thinking he had finally caught him, Frankenstein used his remaining strength to launch a left aimed at Jack's head. Jack ducked and countered with an uppercut that connected with Frankenstein's jaw.

"*Crack.*" Frankenstein's teeth slammed together, and with a dazed look, he passed out and collapsed. The crowd went crazy, with everyone screaming something at once.

When the noise died down, the gunner's mate asked for help getting Frankenstein to sickbay.

Meanwhile, the BB Wonders crowded around Jack, congratulating him.

"Where did you learn how to fight?" said one.

"My Uncle Bill taught me to box, but I really learned how to fight in my Grandfather's railroad yard protecting younger boys from bullies. I hate bullies." *Hitler's a bully*, he thought to himself.

~~~

Saturday, October 19

"J. P., I sure hope Georgia Tech does better against Vanderbilt than they did against Notre Dame last week," Jack said as they looked at the day's football schedule.

"Jack, you might want to check this out. I think you caught someone's eye yesterday."

There in front of them was the usual schedule and assignments. Under Division 6, Jack was listed as a group leader for today's activities.

"Great, more meetings to go to," said Jack, but he smiled.

"Look, Jack, Lt. Commander Bunting is explaining how we get off the boat when the cruise is over. He says left foot first, but I'm thinking I want to start with my right. What do you think?" They both laughed.

Friday, October 25th 1940

Jack and J. P. headed down the gangplank for their much-anticipated leave at Norfolk. Above the din of his fellow seaman, he heard a voice.

"Jack, we're over here."

Jack immediately recognized it as his father's voice, and his heart unexpectedly filled with emotion. He and J. P. turned and saw their fathers standing a few yards away.

"Dad, what are you doing here?" asked Jack, swallowing down the lump forming in his throat.

His father beamed. "We checked with the Navy, and they said we could meet you. We've got two rooms at the Hotel Monticello, and we're planning on taking you two out to dinner. How's that sound?" Frank turned to J. P.'s dad and winked.

"It sounds fantastic," Jack replied, giving his father a hearty embrace. Jack quickly pulled away, embarrassed, realizing that this outward sign of affection was uncharacteristic of their relationship.

Though wide with surprise, Frank's eyes shone.

Jack smiled, a feeling of relief washing over him. "It's great to see you, Dad. Oh, one more thing. Would it be okay if I brought along some of the guys?"

"No problem," said Frank.

Jack turned and whistled, and the three Roanokers — Ed Hall, Frank Long, and Fred Taylor, plus the remaining BB Wonders — came quickly to join them.

Frank and J.P's dad shook their heads and grinned. "Looks like we're having a party," said Frank.

~~~

In short order, Jack, J. P., and the others arrived at the hotel and crowded into Frank Sherertz's room. It was standing room only.

"What do you want to do first?" Frank asked.

"Take a bath," they said in unison, prompting laughter from both dads.

Figure 8. Memo to USS *Arkansas* Apprentice Seamen, (A,B). Jack Sherertz collection. (see page 63)

U.S.S. ARKANSAS

Norfolk, Virginia,
23 October 1940.

MEMO TO ALL PETTY OFFICER SECTION LEADERS OF RESERVISTS AND RESERVISTS.

Subject: Clothing and equipment of V-7 Reservists - Turning in of.

1.   The following schedule for turning in clothing, bedding and other equipment will be followed at the time and places specified below

Thursday afternoon, 1600: The following items will be turned in:
> Bluejackets Manuals
> 1 white hat
> 1 towel
> 1 neckerchief

The 1st, 2nd, 3rd, and 7th divisions will fall in numerically and turn in their gear to the supply department in the starboard aircastle.
The 4th 5th, 6th, and 8th divisions will fall in numerically and turn in their gear to the supply department in the port aircastle.

2.   Transportation home: On Thursday forenoon, Ensign Ashler will publish a list of the men to draw transportation home. In case two or more reservists are going to the same town, one of them will be selected to draw the transportation for the entire group. Only the men whose names appear on Ensign Ashler's list will fall in at 1700, Thursday, to draw transportation. All men from "A" through "K" inclusive will fall in and receive transportation at the starboard side of the navigation study room. Men whose names begin with "L" and inclusive of "Z" will fall in alphabetically and draw transportation from the port side of the navigation study room.

3.   Friday morning: At reveille, all reservists will not lash their hammocks. They will, however, take off the mattress cover and fold it, fold the blanket, take out the hammock clews and place all of the gear in the center of the hammock and then fold the hammock neatly. The reservists will then fall in line in numerical order and turn in hammocks to the port and starboard bunk rooms as follows: Divisions 1, 2, 3, and 7 turn in gear to starboard bunk room. Divisions, 4,5,6 and 8 turn in gear to port bunk room.

4.   Friday morning, 0830: Reservists will fall in numerical order and draw suitcases and civilian clothes. As soon as civilian clothing is drawn, the reservists will shift clothing. When men are shifted, they will fall in numerically and turn in the following articles:
> 1 pea coat
> 1 jersey
> 1 white hat
> 1 towel
> all coat hangers

Locks and keys will also be turned in. The keys must be attached to the locks with string so that keys will not be lost.
> 1 blue jumper
> 1 blue trousers

Divisions 1,2,3, and 7 will turn in articles to the starboard aircastle. Divisions 4,5,6, and 8 will turn in articles to the port aircastle.

5.   From the above it will be noted that reservists will be permitted to keep the underwear and socks issued to them by the government. All other articles must be turned in and no man will be cleared from the ship until he has satisfactorily accounted for all articles issued to him.

6.   The ship expects to moor in the Hudson River, New York, about noon Friday. As soon as all reservists are checked clear they will be disembarked.

S.S. BUNTING.
Lieut.Comdr., U.S.N
Training Officer.

USS *Illinois* (*Prairie State*), Midshipmen's School

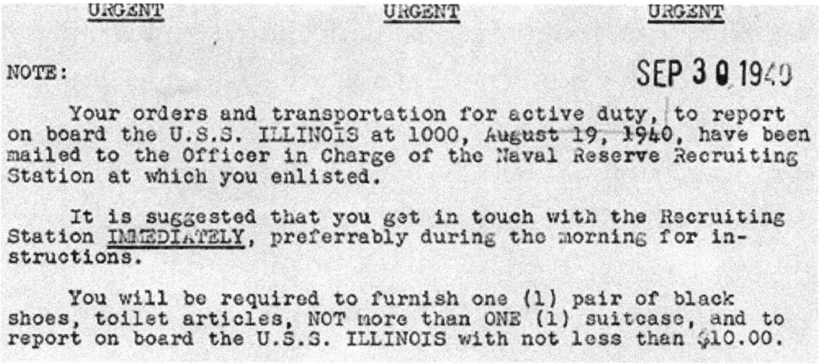

URGENT URGENT URGENT

NOTE: SEP 3 0 1940

    Your orders and transportation for active duty, to report on board the U.S.S. ILLINOIS at 1000, August 19, 1940, have been mailed to the Officer in Charge of the Naval Reserve Recruiting Station at which you enlisted.

    It is suggested that you get in touch with the Recruiting Station IMMEDIATELY, preferrably during the morning for instructions.

    You will be required to furnish one (1) pair of black shoes, toilet articles, NOT more than ONE (1) suitcase, and to report on board the U.S.S. ILLINOIS with not less than $10.00.

Figure 9. Seaman Sherertz's orders to report to the USS *Illinois*.

Figure 10. USS *Illinois*. Jack Sherertz collection.

November 22, 1940, Brooklyn Navy Yard

On Friday, November 22, Jack Sherertz and all of the BB Wonders, except for J. P. Saul, got out of their taxi at West 135th Street and North River and began walking toward the USS *Illinois*. It was mass bedlam ahead, with 1000 white-uniformed midshipmen all arriving at the same time, half came from the USS *Arkansas* and half from the USS *New York*, down from 2000 at the start of their apprentice cruise. As they neared the gangplank, Jack and his friends saw a group of young men in civies, reading the Apprentice Seamen Cruise Itinerary. They stopped for a moment to watch.

"Did we look that terrified?" Jack asked.

The rest nodded and laughed.

Jack felt much better this time, as if he really belonged in the Navy.

As they reached the top of the gangplank, they saw a bulletin board posting recent *New York Times* articles about the War. The USS *Reuben James* had been sunk by a U-boat off the coast of Iceland, with 115 men lost. The British freighter SS Cree was sunk off the coast of Ireland, all 45 men were lost. The 40,000 + ton German battleship *Bismarck* was just reported by German newspapers to have been launched in February. It was the largest ship of its kind in Europe and had 8-15 inch guns.

"Now this is why we're here," said Sherertz, pointing to the article about the Bismarck. "We need to sink the

Bismarck!"

"*Sink the Bismarck!*" echoed several of the BB Wonders.

War is coming fast, thought Jack to himself.

Just below the newspaper clippings were the names of all the men participating in Midshipmen School, divided into three companies. Below that was a schedule for each company's training, with a list of the courses.

Seamanship Instruction Courses

Vessels of war, general characteristics of ships, blinker drill and flag hoist drill, small boats, handling of boats in the surf, rescue work at sea, towing, ground tackle, rope and its uses on board ship, docks, docking, painting, ship handling — use of engines and rudders, ship handling — turning, orders to the wheel, ship handling — mooring, going alongside docks, compass, log, lead, relative bearings, rules of the road (lights and so forth, sound signals for fog, steering and sailing rules, sound signals, distress signals, pilot rules), weather, maneuvering to avoid collision, ship internal organization, duties of the O.O.D. (Officer of the Deck - underway, in formation, at anchor).

Drills

Blinker, visual signals (flags, semaphore, maneuvering ships in obedience to signals), marlinspike—makeup and use of common knots, splicing, seizing, worming, paracelling, mooring and unmooring, naval ceremonies, writing up the log, identification of lights (identification of vessels at night), first aid, damage control, naval law.

"That's a lot of stuff to learn," said Mote.

"That's why they call it school, genius," said Jones.

There were lines everywhere to pick up books, get another vaccination, or get new gear.

"Looks like we're going nowhere fast," said Jack Sherertz.

"Round two," said Rev Drane, as they picked up their duffels and headed down to stow their gear below deck.

Arriving where they would be bunking, they found a bucket next to every bunk.

Drane laughed. "What's this for, swabbing the deck?"

"It's bad news," said Sherertz. "One of the bosun's mates told me they fill it up each morning, and we have to use it for drinking, brushing teeth, and washing clothes."

"Peachy," said Jones.

~~~

One of the few respites during the day from the seemingly endless hours of classroom time was drilling. Even though most midshipmen disliked drilling, Sherertz and Mote loved it. They loved it because of two men in their drill group nicknamed Mutt and Jeff, their class clowns. Mutt was short and Jeff was tall, just like the cartoon characters. To keep drilling from being boring, the two made it into a game and competed to outdo each other each time they drilled.

One particular cold, late afternoon in early December, Mutt and Jeff, Sherertz and Mote, and the rest of their drill group arrived at South Field on the Columbia University campus, having just marched 8 miles from the Brooklyn Navy Yard. Lt. j.g. Reed Adams was addressing them, explaining what they would be expected to do during the graduation ceremony. They were promised if they drilled well, they would be bussed back, but then he hesitated and got very intense.

"Some of you seem to think this is a game," said Adams.

Mutt and Jeff nodded to each other and to the rest of the group as Adams paced away from them.

In response to the resulting snickering, Adams turned and glared. "If I see any cutting up or inappropriate behavior, you will be out of the Navy and the rest of you will be marching back. Are we clear?"

"Yes, sir. Clear, sir," came the yelled response.

"I can't hear you!"

"Clear, sir," came the even louder response.

Sherertz and Mote grinned at each other as they waited to see what would happen next. Mutt and Jeff's game was simple: do something outrageous, sufficient to cause a response in the men around you, but clandestine enough not to get caught.

"All right, then! Let's begin," said Adams. "Attention! Right face, forward, march . . . left . . . left . . . left, right, left."

Jeff put a little skip step in every time he moved his right leg. He did it so fluidly, it looked as if it was part of the drill. Pretty soon some of those around him were trying to imitate it. Mutt, on the other hand, chose to wrap his arms around himself as he marched, as if giving himself a hug. No one imitated him, but the snickering around him got louder and louder. To make matters worse, the pair was right next to each other in formation, amplifying the disruption they were causing.

"To the rear, march!" yelled Adams.

The command had come while one of the nearby men was focused, not on marching, but on Mutt. As a result, he walked straight into the man in front of him, almost knocking him over. Adams had been looking the other way, but the men's laughter brought him running back.

By that time, the company was back in formation, and everyone had a straight face.

"I don't know what happened here, but it better not happen again," Adams yelled. "Column left. Column right. Company halt! That is without doubt the worst display of marching I have ever seen. You are going to keep working on this until you get it right, even if we have to stay here all night."

Almost magically, lights on the field came on, and the drilling continued as day turned into night.

After what seemed like hours, Adams had them get their rifles and return to formation. "Now we'll see what you've learned since last week," he yelled. "Order arms! Right shoulder arms!"

On this command, Jeff arched his back, pushing his stomach out, and Mutt did the reverse.

"Left shoulder arms!"

This time Mutt and Jeff arched in the opposite direction. The men around them increasingly were watching them, rather than paying attention to Adams.

"Present arms!"

Now both Mutt and Jeff began to oscillate in place, almost as if dancing.

"Order arms!"

Although Adams hadn't caught them yet, the direction of the snickers and occasional laughter was slowly narrowing down who might be the cause of the disturbance. Sherertz thought Adams grinned slightly before he started the next series of commands, but he wasn't sure.

"Left shoulder arms!" Adams yelled. "Right shoulder arms! Present arms! Right shoulder arms! Order arms! Left shoulder arms! Right shoulder arms! Left shoulder arms! Present arms!"

Faster and faster came the commands. Disaster was inevitable. Mutt and Jeff kept up for a while, but it was becoming more and more difficult to show off because of the fast pace. Then a man in front of them dropped his rifle, distracting Mutt for a split second as he tried to perform left shoulder arms. Mutt lost his grip on the rifle, and while trying to recover, pushed it right into Sherertz's nose. This started a chain reaction of dropped rifles and others jumping out of the way, until the drill was reduced to total chaos.

~~~

"Order arms!" Adams looked around at the bodies on the ground.

The formation was a shambles. Everyone was scrambling to reassemble in straight lines, and blood poured out of Sherertz's nose.

"Company at ease!" Adams yelled. "Clean yourselves up. We're heading back to base." He waited for them to reassemble and then began again. "Attention . . . Right face . . . Forward march."

Adams marched double time up to the front of the column, and when his back was to the group and his face couldn't be seen, a big smile appeared. "I didn't catch you this time, but I made you pay, and now you're marching home," he said. "All right gentlemen, follow me." He got into a waiting jeep and immediately went into a singsong marching song, which the company repeated back.

> They say that in the Navy, the coffee's mighty fine,
>
> Looks like muddy water, tastes like turpentine.
>
> Oh Lord, I want to go home,
>
> But they won't let me go home, o-o-o-o-o-oh hey!

> They say that in the Navy, the chicken's mighty fine,
>
> One jumped on the table, and started double-time.
>
> Oh Lord, I want to go home,
>
> But they won't let me go home, o-o-o-o-o-oh hey!

> They say that in the Navy, the fish is mighty fine,
>
> One jumped out the water, and killed a friend of mine.

Oh Lord. I want to go home,

But they won't let me go home, o-o-o-o-o-
oh hey!

They say that in the Navy, the pay is
mighty fine,

They give you hundred dollars and take
back ninety-nine.

Oh Lord. I want to go home,

But they won't let me go home, o-o-o-o-o-
oh hey!

~~~

The next night, it was time for another Thursday night dancing class. This time there were about forty women, up from fifteen in the first couple of classes, brought in to try and help five hundred midshipmen learn to dance. A fifteen-piece dance band started playing, and almost all of the men suddenly looked awkward. A few, like Sherertz, headed straight over to the women and asked them to dance. The rest of the BB Wonders just stood there, shaking their heads.

"Can you believe Sherertz?" asked Tommy Jones. "He got hit in the nose with a rifle so he looks like he's been in a bar fight, and still he gets the women."

"Damn," said Johnny Mote. "I guess I'm going to have to take the plunge." Off he went to ask the last unattached girl to dance.

The rest of the BB Wonders looked on until Sherertz walked back over to them.

"Look, Sherertz, this is embarrassing," said Jones. "Can you teach us how to dance?"

Sherertz laughed. "Who's first?"

"I am," said Jones in a high voice. He took his handkerchief out of his pocket and tied it around his neck like a woman's scarf. "How do I look, Jack?"

"Just gorgeous," said Jack. "Shall we dance?" He held out his arm to escort his partner onto the floor. After several turns around the floor, Jack brought Jones back to find that the other BB Wonders had been primping in his absence. Several had rearranged their hair, one had taken off his coat and tie and unbuttoned the top button of his shirt, and one had put on lipstick. "Who's next, my beauties?"

"Pick me. No me," came a chorus of high-pitched voices. Pretty soon their whole end of the dance floor was laughing hysterically.

~~~

December 20, 1940

It was Friday night, and the BB Wonders had just arrived at New York City's Stork Club. They planned to head home for Christmas the next day. They were met at the door by the maître d', all decked out in a tux, who verified their reservation and motioned for them to follow him. He led them through several rooms until they arrived at a larger room where a jazz band was playing a hot number. Two tables had been pulled together and covered with white tablecloths that hung to the floor. Their tables were located down front next to the band and the dance floor. The walls were covered with beautiful dark gray fabric, punctuated by gold metal light sconces alternating with mirrors, that made the place look like it went on forever. The ceiling lights were quite modern, with no chandeliers to be seen. The dance floor was full, and loud music made it hard to hear anyone speak.

Jack Sherertz spun around several times next to the table, holding an imaginary dance partner. "This is my kind of place," he said.

"I hope so, since you picked it out," said Johnny Mote. "How did you get us down front?"

"A big tip works wonders," said Sherertz.

"Hey, Rev, who's your buddy?" asked Tommy Mote.

"Guys, I want you to meet Charley Sterns, from Albany, New York. He went to Dartmouth, finished up in '36, and for the last four years has been on a quest to find the perfect woman. He signed up about the same time we did and took his cruise on the USS *New York*, that big hulk that shadowed us down to the Caribbean and back. He joined us here on the *Illinois*."

"It is not a quest, as I have found my fair damsel," said Sterns. "Meet Miss Joan Mayer from Berkeley Hills,

California." He brought his girl forward with a flourish and leaned forward to kiss her hand. She turned red but beamed back at him. It was pretty obvious they were in love.

"Oh, and by the way, he's a funny guy," said Rev Drane. "*Writes for the Side Boy,* so be careful what you say. And one more thing: he also has seen the light and is battleship bound."

"We welcome another enlightened spirit into the inner circle of wisdom," said Johnny Mote.

"Hey, did you guys hear about all the guys in our class that washed out because they flunked the eye test or couldn't do push-ups?" Garfield asked. "They're dropping like flies."

Sherertz began hitting his water glass with his fork. "I would like to make a toast," he said as he stood and faced his friends. "To J. P. Saul, Rabbit and all our other shipmates who can't be here with us tonight."

"To Jack, the Frankenstein killer," said Jones, prompting a laugh from everyone.

"To calm seas and safe harbors and good friends always," said Jack, and they all drank their champagne.

"Enough of that sappy talk," said Mote. "I came here to have a good time."

~~~

One Saturday night, Midshipman Sherertz was standing the second watch (2000–2400) as the junior officer of the deck. He had on his service dress whites with a sword belt. His job was to guard the gangplank leading from the shore to the USS *Illinois* and challenge all those seeking to come aboard. Aside from the chilling cold, the night was uneventful until midshipmen started returning from shore leave. There was a strict rule against becoming drunk and disorderly, but it was blatantly ignored as long as the midshipmen did not cause a ruckus that brought their drinking to the attention of their senior officers. Midshipmen had to be back by 10:00 p.m., and all those on leave typically pushed this time to the maximum so that everyone seemed to return at once. This resulted in a steady stream of men returning, each requiring a challenge from Sherertz. In the gaps between groups, Sherertz could see some of the men kissing their girlfriends goodnight in the adjacent parking lot.

Then Sherertz saw some of his buddies returning. It was obvious that they had all been drinking, but Tommy Jones looked the worst. He was being supported in between Drane and Mote.

"Who goes there?" challenged Sherertz.

"A bunch of drunk sailors," said Uhrenholdt, walking in front of the other three.

"State your business," said Sherertz.

"Cut the crap, Sherertz," said Uhrenholdt. "We need to get Tommy to bed before someone sees him. No way he can pass for sober."

With that, Tommy let out what sounded like a banshee wail that changed at the end to whimpering.

"What's his problem, Andy?" Sherertz asked quietly.

"He got a 'Dear Tommy' letter from his girlfriend this afternoon—or I should say ex-girlfriend, and he's been torn up ever since."

"Get him below quickly, Andy, before you get us all thrown out of the Navy," said Sherertz.

"Is that you, Jack?" asked Tommy in a very loud, drunk voice. "I need a date. She dumped me."

Andy, Drane, and Mote picked him up, ran up the gangplank, and disappeared down a hatch to their bunks below. Drane, who had him under the arms, clamped one hand over his mouth. A good thing, as another wail came out, barely damped by Drane's hand.

A moment later, a loud voice from an open hatch yelled, "What's all the commotion out there, Sherertz?"

"No problem, sir," replied Sherertz. "Just a bunch of rowdy townies trying to give our men a hard time. We took care of it."

Shortly after 10:00 p.m., it became dead quiet again. Minutes later, Sherertz could hear crunching sounds coming from the parking lot. They seemed to get louder. While no one really suspected that there were spies, they were taught to be watchful just in case.

"Who goes there?" challenged Sherertz in a loud voice.

There was no response, but the sounds seemed to get closer.

"Who goes there?" he repeated.

Again, no response.

Finally, Sherertz walked down the gangplank to investigate. As he did, he took out a bosun's pipe and blew the note that was supposed to alert the officer of the deck to a possible problem.

A moment later, the OD called down from the deck above, "What's the problem, Sherertz?"

"Listen," said Sherertz.

The crunching noises were quite apparent.

"I'm going to investigate," said Sherertz.

"I'll guard your post," said the OD.

Attempting to be quiet, Sherertz walked slowly down the rest of the gangplank and then into the parking lot, toward the crunching noise. In the still, cold night, it sounded quite loud. Not knowing what to expect, he steeled himself for a possible confrontation with some dangerous threat. Finally, it was clear that the noise was coming from between two parked cars. He turned to face the danger, only to see a rat chewing on a piece of frozen meat. Sherertz stomped his foot, and the rat took off. Without saying a word, Sherertz walked back up the gangway with a serious look on his face.

The OD looked worried. "What was it? Do I need to

wake up the brass or call the cops?"

Totally serious, Sherertz said very slowly, "It was a giant, vicious, man-eating . . . rat." He broke out laughing, to the displeasure of the OD, who realized he had just been had.

~~~

February 8, 1941

Today Sherertz and Mote and Ehrenholdt would go on their *Sylph* cruise. The *Sylph* was a yacht built in 1929 by George Lawley & Sons and then given to the Navy at the beginning of World War II. It was originally called Intrepid and then renamed *Sylph* (*PY-12*) by the Navy. It was repurposed to train midshipmen and launched in October 1940.

Sherertz was excited about the cruise because it got them out of class for the day, and they would be able to shoot guns, instead of merely talking about them. When their group of forty midshipmen assembled for their final instructions they were told they would be cruising up the Hudson River while practicing their seamanship skills. On the way back, they would get to use their rifles to shoot at targets towed behind them in the water. And one more thing, the captain told them, it was 15 degrees and snowing.

By the time they finished going up the river against the wind, they all looked miserable, with one exception: Andy Uhrenholdt. The only one on the cruise not shivering violently, he kept mumbling something about hunting weather in Wisconsin and seemed to be enjoying himself thoroughly. When they came about to go back down the river, they were sailing with the wind, so it felt a little less cold — but only a little.

A bosun's mate came out and announced what they had all been waiting to hear. "Target practice will commence immediately. Secure your rifles."

Everyone went to the weapons locker, where they were issued an M1 Garand rifle and ammunition, and then to the stern of the boat. A wood box attached to the *Sylph* by a rope bounced around in their wake about fifty yards behind the boat. One by one, they tried to hit the target. Between their continued shivering and cold hands, no one was coming close. Then it was Uhrenholdt's turn. He walked up, sighted carefully, and hit the box with his first shot.

"Nice shootin', sailor," said the bosun's mate. "Care to try again."

"Don't mind if I do."

He proceeded to shoot six more shots into the wooden box, until he reduced it to wood planks that no longer resembled a box. "What's next?" he asked with a grin.

~~~

Finally the forty midshipmen docked and disembarked down the gangplank, all in a hurry to get warm. As they headed up the gangplank to the *Prairie State*, Sherertz yelled, "Don't forget tonight. We need to rank our ships."

~~~

The BB Wonders had gathered in the back of a bar, not far from the USS *Illinois*, to talk over something important. They each had a mug of beer, and two pitchers for refills. They looked like they planned to be there awhile.

"Well, gentlemen, it's decision time," said Jack Sherertz. "Tomorrow we have to turn in our rankings for which ship we want to serve on."

"And don't forget, tomorrow we also have our weekly seamanship quiz," said Tommy Jones.

"Aaah, shut up, Jones," said Johnny Mote. " This is more important."

"Shut up yourself," said Jones.

"Moving right along," said Jerry Garfield. "It looks like we shouldn't have any problem serving on a battleship, since so many of our classmates want to serve on carriers."

"Who would want to serve on a carrier, unless you were a pilot?" asked Charley Stern.

"I don't know, but some of the officers think that carriers are going to be the next big thing," said Jones.

"Who said that?" asked Uhrenholdt.

"Well, it was Lieutenant de Forest and Ensign Hamilton," said Jones.

"Well, there you go," said Uhrenholdt.

"What does that mean?" Jones asked.

"Well, de Forest is a chaplain, and Hamilton only has a little more experience than we do," said Uhrenholdt.

"Still . . ." said Jones.

"Well, I'm still going with battleships," said Rev Drane. "Shall we go over our options?" He looked around.

Everyone, including Jones, nodded.

"Who's got the list?" asked Garfield.

"I do," said Mote. "Here are the choices. Division 1: *Arizona, Nevada,* and *Pennsylvania.* Division 2: *Tennessee, Oklahoma,* and *California.* Division 3: *Idaho, Mississippi,* and *New Mexico.* Division 4: *West Virginia, Colorado,* and *Maryland.* Division 5: *Texas, Arkansas,* and *New York.*"

"So what do we know about them?" asked Sherertz.

"They're more similar than different," said Uhrenholdt. "They were all built just before, during, or just after World War I. They range from twenty-five thousand to thirty-three thousand tons, they all do twenty to twenty-one knots, and they all shoot torpedoes except the *California.*"

"All right, so what's different that will help us decide?" asked Garfield.

"The number of men on board varies quite a bit. It ranges from eight hundred sixty on the low end all the way up to over fourteen hundred on the *West Virginia,*" said Stern.

"That doesn't help me very much," said Sherertz.

"I looked up armament," said Jones. "The big guns

range from twelve inches on the *Arkansas*, up to sixteen inches on the *Colorado*, *Maryland*, and *West Virginia*. Most have fourteen-inch guns. The five-inch guns are pretty similar, but range in number from twelve to thirty. The big differences seem to be with the antiaircraft guns. They range from none on the *Nevada*, *Oklahoma*, *New Mexico*, and *New York*, to anywhere from two to eight three-inch guns on most of them, to large numbers of 40 mm and 20 mm guns on the *California* and *West V*irginia. The last two have the most maneuverable guns. So if you believe air power is going to be the cat's meow, I give the advantage to the *California* and *West Virginia*, or at least to those with some antiaircraft capacity. I've summarized it all in this table." He passed the table around.

"Wow, this is amazing," said Drane.

The rest nodded, as they looked it over.

"There's still one thing we've been avoiding, that may be the crux of the whole matter," said Sherertz. "All the action seems to be in the Atlantic, yet most of the battleships seem to be in the Pacific."

"I overheard Captain London telling one of the other officers that he thinks the German navy isn't far from being out of business at the rate the Brits are sinking their ships," said Mote.

"Yeah, but the Atlantic convoys still need protection," said Jones. "Just since we started school, German ships have sunk several merchant ships and a British cruiser."

"How will a battleship stand up against the German submarine packs?" Uhrenholdt asked.

"It's not their job; that's for the destroyers," said Garfield.

"Has anyone heard anything about Admiral Kimmel, the new top dog in the Pacific?" Stern asked.

"Just scuttlebutt," Sherertz answered. "Nothing helpful."

They went back and forth for hours debating everything they knew about battleships.

"I think we've skinned this cat about every way possible," said Garfield.

They all nodded and stood up to leave.

"Good luck, gents," said Sherertz. "I hope everyone gets what they want. Listen, I have an idea about how we can keep in touch after graduation. How about a Friday night poker game when any of us are in the same town?"

"Great idea, Sherertz," said several. Everyone nodded enthusiastically and shook hands. Then the group adjourned.

~~~

February 28th, 1941, V-7 Graduation

At the start of their graduation ceremony, Jack Sherertz and the other 468 midshipmen in the second graduating V-7 class marched onto the football field of Columbia University, accompanied by a Navy band. When they arrived on the field, they stopped at the sound of "Brigade halt." Following more commands, each line of midshipmen filed down a row of chairs until they reached the end, halted in front of their chairs, and did a right face, where they waited at attention. When all of the midshipmen were in position, the command "Be seated" was issued.

The midshipmen faced an elevated platform arranged to seat all their instructors, local dignitaries, and visiting Navy brass. In the background were the majestic columns of the main building of Columbia University. Parents and friends were seated in the bleachers surrounding the football field.

As Jack sat there, he was able to locate his parents. He was quite excited, and glad they were there to share this with him.

Finally the ceremony began with Captain John London, Commanding Officer of the US Naval Reserve Midshipmen School presiding.

"Congratulations to the second V-7 graduating class. You have worked hard and we welcome you into the Navy. The need is great. German U-boats are sinking British ships at a furious rate. During the three months that you completed your training 15 British warships were sunk, including one battleship and one aircraft carrier. Just five days ago a British freighter flying US colors and with a US flag painted on its hull was sunk in the Indian Ocean. I can't tell you when the US will join the war, but I think it likely will be soon. Good luck and Godspeed.

Thereafter, a seemingly endless series of people talked

about the importance of being ready for the war, being ready to support our allies, training the best and the brightest to be ready to face all the challenges, and so on. At last all of the midshipmen stood to hear their names read in alphabetical order, along with their next assignment. Jack listened to each one so he wouldn't miss any of his friends.

"Walter H. Drane, Ensign USNR, USS *Nevada*."

Jack got very excited when he heard the first of his friends designated an ensign.

The rest followed: "Jerome H. Garfield, Ensign USNR, USS *Arizona* . . . Thomas R. Jones, Ensign USNR, USS *Arizona* . . . John L. Mote, Ensign USNR, USS *Nevada* . . ." and then he heard it: "Herbert J. Sherertz, Ensign USNR, USS *Nevada*."

He couldn't believe it; he would serve with two of his classmate friends on the same battleship.

The roll call continued: "Charles M. Sterns Jr., USS *Oklahoma* . . . Andrew C. Uhrenholdt, USS *Arizona*."

Finally, "Congratulations to all new ensigns of the class of February 1941." By the end of the war the V-7 program would train more than 50,000 ensigns, over and above the 18,000 naval officers in place before this training started.

The ensigns threw their hats into the air and cheered as the Naval Academy band played "Anchors Away." Afterward they were free to go find their families, and Jack headed toward the stadium to find his parents.

How amazing, he thought. The BB Wonders had all been

assigned to battleships and would all be together at Pearl Harbor. What could be better than that?

U. S. NAVAL RESERVE MIDSHIPMEN'S SCHOOL
on board U.S.S.ILLINOIS, New York, N.Y.

SEAMANSHIP DEPARTMENT INSTRUCTION SCHEDULE

SECOND MONTH

DATE	NO	TOPIC	TEXT	ASSIGNMENT
Monday, Dec.23	19	COMPASS, LOG, LEAD, RELATIVE BEARINGS	Knight	Chap.VI, p 124-165
Tuesday Dec.24 Thurs. Jan.2'41	20	RULES OF THE ROAD; enacting clause, scope,etc.Lights, etc.	Knight	Chap.XII thru Art.8, p.367 READ ALL FOOTNOTES
Friday, Jan.3	21	RULES OF THE ROAD; lights and so forth continued	Knight	Chap.XII Art.9(p.369) thru Art.14(p.378) READ ALL FOOTNOTES
Monday, Jan.6	22	RULES OF THE ROAD; sound signals for fog	Knight	Chap.XII Art.15(p.379) to end of Art.16, p.396. See also table p.443 READ ALL FOOTNOTES
Tuesday, Jan.7	23	RULES OF THE ROAD; Steering and sailing rules	Knight	Chap.XII p.394-5 thru Art.27, p.428-9 READ ALL FOOTNOTES
Wed'day, Jan.8	24	RULES OF THE ROAD Sound signals, distress signals,etc. WEEKLY QUIZ	Knight	Chap.XII Art.28 (p.430-] thru Art.32 READ ALL FOOTNOTES
Thurs., Jan.9	25	RULES OF THE ROAD; Pilot Rules	Knight	Chap.XII, p.459-471, incl. READ ALL FOOTNOTES
Friday, Jan.10	26	WEATHER	Knight	Chap.XX, STUDY Parts III VI, VIII, X & IX. READ Parts I,II,IV & V
Monday Jan.13	27	MANEUVERING TO AVOID COLLISION	Knight	Chap.XIII (p.472) as far as "Collision Mats"
Tuesday, Jan.14	28	SHIP INTERNAL ORGANIZATION;Duties and relationships of officers	N.Ad.I	A-133 thru B-209; omit A-134
Wed'day, Jan.15	29	SHIP INTERNAL ORGANIZATION; organization of crew; gen'l drills WEEKLY QUIZ	N.Ad;I B.J.M.	A-134; B-210 Chap.26
Thurs., Jan.16	30	DUTIES OF THE O.O.D. UNDERWAY	N.Ad.I W.O.G.	B-226 to B-264 Chaps.I to IV,incl.
Friday, Jan.17	31	DUTIES OF THE O.O.D. UNDERWAY	W.O.G.	Chaps. XXI & XXIII Omit foreign signals

Figure 11. US Naval Reserve Midshipmen's School Seamanship Schedule.
Jack Sherertz collection.

89

Figure 12. BB Wonders Ensign Graduation pictures. *Side Boy*, February 1941.
Can be found at www.AHeroAmongMillions.com.

Figure 13. Navy Midshipmen, Columbia University, South Field, 1942.
Columbia University Archives (Gift of Charles A. Riddle III).
Can be found at www.college.columbia.edu/cct/summer13/columbia_forum
or at www.AHeroAmongMillions.com.

Pre-war Pearl Harbor

Figure 14. Pearl Harbor, October 30, 1941. Ford Island in center. Battleship Row just to left of Ford Island. US Navy National Archives, history.navy.mil, Museum (Naval Aviation Photo # 80-G-182874, Wikipedia).

April 4, 1941

Ensigns Sherertz, Mote, and Drane stood up for the entire ride on the whaler that transported them from the officer's dock to their first arrival at the USS *Nevada*. The USS *Nevada* was a battleship launched in 1914, the first of the super-dreadnought class characterized by triple gun turrets, burning oil to generate steam, and "all or nothing" armor. It participated in World War I and was now one of the core battleships in the Pacific fleet. To the three green ensigns, it was a thing of beauty.

Ensigns Sherertz, Mote, and Drane were dropped off on the portside gangway of the USS Nevada with their duffel bags. They headed up to the deck, where they found themselves staring at the forward fourteen-inch guns and the various observation decks in the towers just aft. Like deer in the headlights, they kept staring until they became aware of someone standing near them.

"Welcome to the USS *Nevada*, Ensigns. I'm Captain Scanlon, and this is Chief Bosun's Mate Hill. He'll get you started."

They all saluted in unison.

"At ease, men. I'll leave you in Chief Hill's capable hands," said Scanlon as he walked away.

"Stow your gear in Junior Officers' Country, reachable via that portal over there on the second level below deck." Chief Hill pointed in the direction they were to go. "You'll see your names above the entry to your berth. Then come right back up here, and I'll give you your initial assignments."

"Aye, aye, sir," said the ensigns.

Sherertz and the others went flying through the indicated portal, threw their duffels down first, and slid down the first set of rails and then the second set of rails. They hit the second deck, clueless about where to go next.

Then along came a seaman who looked as green as they felt.

"Seaman, where's Junior Officers' Country?" Sherertz asked.

"Forward through that portal, sir," said the seaman. He pointed toward a portal in a bulkhead. Above the portal, a sign said, *Warrant Officers' Country*.

They stepped through the portal and started walking. On the right they passed signs reading, *Warrant Officers' Pantry*, *Junior Officers' Pantry*, and *Junior Officers' Mess Room*. Finally they reached another bulkhead with a portal. Above the portal, it read, *Junior Officers' Country*.

Sherertz felt a chill go up and down his spine. They had reached their new home. Stepping through the second portal, they encountered a series of doors on both sides of the ship, extending forward toward the bow end of the boat. They headed toward the port side and saw that each door had two ensigns' names above it. After passing several doors, they came to a sign that said, *Ensign Drane*, Ensign Sherertz. They were assigned to room together because they were both deck officers. They turned the handle on the door and checked out the cabin. On each side was a bed attached to the wall, with storage above it. There was a small desk for each of them. Sherertz and Drane each found a locker with their name on it and stowed their duffels inside. They would have plenty of time to unpack later. As they came out, they saw Mote exiting his quarters, and they all headed back up to the main deck.

Exiting the portal, they found a seaman and asked the location of Chief Hill.

"Sirs, usually you can merely listen for his voice, but I don't hear it at the moment. Why don't you try over near the catapults on the starboard side? I think he's breaking in a brand new group of sea dogs."

"Thanks, Seaman," said Ensign Drane.

They walked forward slightly, and then Sherertz turned right under the aft fourteen-inch gun battery. *Wow,* he thought, as he reached up toward the gun barrels. *We can do some damage with these things.*

Still mesmerized by the guns, he backed out from under them toward the starboard side. Slowly turning, he saw something that took his breath away—the mighty USS *Arizona*, their sister ship. Planks had been placed so that you could walk from one deck to another. Then looking forward of their bow, he saw the USS *Tennessee* and the USS *West Virginia*, and in front of them, the USS *Maryland* and the USS *Oklahoma*.

"Mine eyes, have you ever seen a finer sight?" said Sherertz, mimicking an Irish accent. "I think we're going to enjoy our stay on these boats. I can't wait to see Jones's and Garfield's quarters."

Then they heard an unmistakable sound—a group of seamen holystoning the deck. Chief Hill stood right next to them, a stern look on his face. Sherertz, Drane, and Mote walked up near him.

Jack smiled. "Some things are the same on all ships."

"Well, what do we have here?" Chief Hill asked loudly, for the enlisted men's benefit. "A bunch of green officers. Excuse me, sea dogs, I think I have to break a few stallions. Sherertz and Drane, you stay with me. Mote, you report to Ensign James Egan. You should find him below at a briefing in the Junior Officers' Ward Room."

"Aye, aye, sir," said Mote, and he walked away.

"Sherertz and Drane, you're in charge of our observation crew," said Chief Hill. "Your duties will include oversight of all catapult launches, keeping the lookouts on task, keeping the boat at proper speed, and keeping the helmsman on course. If you'll follow me, I'll begin by showing you how our catapult works."

Chief Hill had started going through the nuts and bolts of how to fire the catapult, when up came another ensign with a set of wings on his lapel.

"Perfect timing," said Chief Hill. "Ensign Davis is one of our pilots. I'll let him finish the catapult tour. I'll show you the rest of your duties later."

"Welcome to the *Nevada*. My name is Curt Davis, from Rock County, Wisconsin."

"I'm Jack Sherertz from Roanoke, Virginia, and this is Rev Drane from Cleveland Heights, Ohio."

As they stood there talking, two other ensigns walked by. Ensign Sherertz turned to introduce himself, but they ignored him.

"Don't bother," said Ensign Davis. "They're Annapolis grads. They think they're too good for us. I've even heard their shit don't smell."

Sherertz and Drane broke out laughing. "I like this guy," said Sherertz.

"There's even a pecking order among Naval Reserve ensigns," said Davis. "Those on battleships seem to think they're better than anyone else, including those on carriers. Don't bother trying to talk with the enlisted men, either. They've been terrorized by the Annapolis officers into believing that's inappropriate."

"Who do you hang out with when you're on leave?" asked Sherertz.

"I haven't made many friends yet," said Davis.

Sherertz and Drane shared a nod.

"Well, Jack and I would like to invite you to join our fellow V-7 ensigns in our Friday night poker game," Drane said.

"I'd like that."

"So what's the basic schedule around here?" Sherertz asked.

"Reveille is at five thirty," Davis answered. "You two oversee the observer crew each time we go out on maneuvers. There are three battleship divisions based here at Pearl. Two are out on maneuvers at any given time for seven- to fourteen-day trips. When you get back, everyone gets leave. Enlisted men get 'Cinderella leave,' which means back by midnight. Officers can stay overnight, with the permission of their senior officer. And, if you're so inclined, each battleship has athletic teams—football, baseball, boxing, and rowing. We also have a band. They're pretty damn good. Oh, and they play 'The Star-Spangled Banner' every morning at zero eight hundred hours, and I do mean every morning."

"What's the next mission?" Drane asked.

"Tomorrow we head out to do some target practice against barges towed by destroyers," said Davis. "I get to fly observation and tell you how badly you miss." He laughed.

~~~

May 1941

"When do you think we'll get some real action?" said Sherertz, as Mote stepped on the launch with Sherertz and some other junior officers for a night of shore leave. "The Germans are still chewing up Atlantic convoys, while it looks like all we'll be doing is drilling."

"Why don't you ask Admiral Kimmel, you know, CINCPAC (Commander in Chief of the Pacific), or better yet, FDR?" said Davis with a wink. "I'm sure they would both be interested in what a green Ensign has to say."

"Yeah, right," said Sherertz. "Heh, Mote. Did you hear that the Brits sank the Bismarck?"

"What's the *Bismarck*?" one of the other officers asked.

"One of the biggest battleship ever built. The Germans

launched her nine months ago for active duty. She was supposed to rule the seas, but the Brits tracked her down in the North Atlantic and sank her in a three-day epic battle. Lt. Commander Thomas said they just got the details and told me the story. The British battleships *Hood* and Prince *of Wales* caught up with her near Greenland and started firing at her from a distance of twelve miles. When the *Bismarck* returned fire, she sank the Hood. Over fourteen hundred men died."

"How can you sink a ship from twelve miles away?" another asked.

By this time Jack realized he had an audience. "The *Bismarck* fired fifteen-inch shells from guns that are sixty feet long," said Jack.

"No shit," said another, holding up his hands and trying to imagine a shell bigger than the fourteen-inch shells on their boat.

"The *Bismarck* suffered only minor damage, but the Germans decided to make a run for it because they thought other big ships might be nearby," said Jack. "The Brits sent all their ships in the area after her. They finally did serious damage, using torpedo planes, disabling her rudder so she couldn't maneuver. Then two British battleships and some destroyers pounded her with shells, taking out all of her guns, and finally sank her."

"Torpedo planes! Do you think they could attack our battleships here at Pearl?" said Mote.

"No idea. We'll have to ask the skipper that question," said Sherertz.

~~~

September 1941

Sherertz, Mote, and Drane sat at a table in JO Country.

"Well, Mote, I'm beginning to think our poker game isn't going to happen tonight," said Sherertz. "Is the *Arizona* having a bake sale?"

"It's even worse than that," said Mote. "Uhrenholdt said their skipper scheduled another emergency meeting."

"And before you think we got off easy, we've got one in an hour," said Drane.

"Well, it's no surprise, with all the shit going on around us," said Mote. "It was one thing when Roosevelt embargoed all US oil exports to Japan, but when he seized all Japanese assets in the US, all our top brass really flipped out. We've been running drills round the clock ever since."

"What's the latest meeting about?" asked Drane.

"Nobody knows for sure," said Sherertz. "Rumor has it that it relates to the USS *Greer* having a shoot-out with a German sub last week. FDR even did a fireside chat about it."

"Sumpin's going to happen," said Mote. "I feel it in my bones."

"You ain't just a whistling Dixie," said Sherertz. "But something else concerns me. No sleep, no leave. Someone is going to screw up."

~~~

October 1941

Division 1 and Division 2 battleships were on maneuvers at the end of a long stint at sea. Division 1 comprised the USS *Arizona*, the USS *Nevada*, and the USS *Pennsylvania*. Division 2 comprised the USS *Oklahoma*, the USS *Tennessee*, and the USS *California*. They were practicing a maneuver designed to protect carriers from a flank attack by submarines or torpedo planes. Instead of having the carriers trail the battleships, they moved to a position surrounded by battleships – both still ringed by destroyers and destroyer escorts. It meant moving battleships quite close to each other and required difficult maneuvering, especially if the seas were the least bit rough. Everyone was exhausted.

Drane and Sherertz had their full observer group manning their posts all over the *Nevada*.

As Sherertz walked around from post to post, one of the seaman called to him. "Ensign Sherertz, doesn't it look like the *Oklahoma* is getting too close to the *Arizona*?"

"Damn," said Sherertz.

Without another word, he went running to a phone to call the bridge to warn of the impending collision.

"Crash," he heard, as he reached for the phone.

The *Arizona* had been rammed by the *Oklahoma*. When the ships separated, it was obvious that the damage was only cosmetic.

Heads will roll, thought Sherertz.

~~~

December 5, 1941, Morning

Figure 15. US Pacific Fleet, April 1940, Lahaina, Maui. From left: USS Lexington (CV 2), USS California (BB 44), USS Nevada (BB 36) & USS Pennsylvania (BB 38). Photograph taken by Floyd Proffitt. Can be seen at www.navsource.org/ archives/01/38a.htm or at www.AHeroAmongMillions.

Battleship Division 1, minus the USS *Pennsylvania*, which was back at Pearl in dry dock, and Division 2 had just anchored off Lahaina on the west coast of Maui after being out almost two weeks. They were stopping briefly while some other ships cleared the harbor at Pearl, and then they were heading in for refueling and to give the men some well-deserved shore leave. Sherertz and Drane were in the forward five-inch antiaircraft director of the USS Nevada, training their 7x50 binoculars out to sea and keeping an eye on the twenty-plus spotters under their command.

"Hey, Rev," said Sherertz, "those damn Jap subs are still hanging around. Our spotters have been tracking them for two days."

"I wish the skipper would green light us to use those subs for target practice, rather than the old scows and barges the destroyers have been towing," said Drane. "Jack, have our boys watch them closely as we head toward

Oahu. If they don't peel off like they usually do, then it will be open season on sinking them. How many are following us this time?"

"There are five that they recognize, following us in rotation," said Jack. "They've named them after the Marx Brothers. Groucho and Chico are following us at the moment."

"Jack, tell them to make sure that the Marx Brothers don't get away with any *Monkey Business*."

"That's a good one, Rev. We'll make *Duck Soup* out of them if they do."

This time Drane laughed. "Jack, look. Our football team is still practicing there on the foredeck, even though Admiral Kidd pulled the plug on the season."

"That sure started the rumors flying," said Jack. "Heard any more scuttlebutt about what's going on?"

"No. I hope we'll hear something tonight at the poker game."

Just then a piercing whistle came over the PA system, announcing they were pulling up anchor and heading to Pearl.

~~~

Battleships steaming into Pearl Harbor were a stirring sight, causing all those in Pearl Harbor not busy doing something else to stop and watch. They steamed single file past Hospital Point, and then turned to port past Ford Island, swinging north around the island and then back down the channel so their bows would be facing the harbor entrance. The *California* went first and was quickly secured to its quays near the ferry landing. Then came the *Oklahoma*, which was anchored channel side, next to the *Maryland*. The *Tennessee* followed next. Since the USS *West Virginia* was momentarily out on a short shakedown cruise, the USS *Tennessee* took its place just behind the *Maryland*, next to two mooring quays. Finally, the *Arizona* was anchored to its mooring quays situated just behind the *Tennessee*, and the *Nevada* was anchored channel side to the *Arizona*. As soon as all anchors and mooring lines were secure, wood planks were placed across from the *Maryland* to the *Oklahoma* and from the *Nevada* to the *Arizona*, allowing easy access back and forth and making them sister ships.

USS Arizona Poker Game

December 5th, 1941, Evening

The BB Wonders were assembled in the junior officers' stateroom on the second level of the bow end of the USS *Arizona* for their Friday night at anchor poker game. The room was full of smoke, and the conversation was lively.

"So Sherertz, what's the matter with Georgia Tech football the last two years?" asked Ensign Andy Uhrenholdt. "Georgia sure kicked your butts last weekend."

"Same thing that's wrong with your Badgers, Andy. We can't seem to find any football players. I think they all signed up for the war."

"At least there's one team we can all take pride in," said Rev Drane. "Go, Navy! Hoorah!"

"Hoorah!" everyone responded.

"Besides, we have right here among us a genuine Naval Academy graduate," said Drane. "Check out his ring." He held up Ensign Joe Taussig's hand.

"What's the matter, Taussig, you couldn't find any blueblood Navy functions to go to, or are you just slumming?" asked Ensign Robert Booth.

"I have to put up with enough of that crap going to my

dad's parties," said Taussig. "I prefer hanging out with you riffraff reservists."

Everyone cheered.

"And besides, Booth, I plan on personally seeing to it that the Crab Bowl Classic is reinstated, so we can kick the U of Maryland's ass on the football field."

"Who's your dad?" asked Ensign Charles Stern.

"Just one of those admiral kind of guys," said Drane.

"Ooooh," said Stern. "Sorry I asked."

Jones came into the room dragging a wide-eyed, nervous-looking ensign behind him. "I want you guys to meet John Morrison. He arrived in Pearl just in time to go on our last cruise. He's being groomed to oversee one of the port gun turrets. Supposedly he's a card shark, so watch out."

"Hi, guys, thanks for the invite," said Morrison. "I promise I won't win all the money."

"First things first," said Garfield. "Let's see the color of your money."

"For you latecomers, the other new guy to the group tonight is Lt. Frank Samuels, one of the docs at the hospital," said Sherertz. "He's been letting me make rounds with him and also claims he's a good poker player."

"Okay by me," said Garfield. "That just means bigger pots for me to win."

"Too bad about Captain Foy," said Garfield. "He

sounded like a good guy, as captains go. From what Sherertz and Drane said, anyone could have bumped the *Oklahoma* into the *Arizona* during that maneuver. Charlie, how's your new captain on the *Oklahoma* working out?"

"His name is Howard Bode," said Stern. "He's a real tightass who doesn't speak to JOs like us. So nobody likes him."

"So, Davis, see anything interesting, like Jap airplanes, while you were flying your scouting runs last week?" asked Mote.

"Nothing but a few gooney birds," said Ensign Curt Davis.

"Say, what's happening this weekend?" asked Mote. "I need some action."

"I've got a table reserved at the Royal Hawaiian for the big dance," said Sherertz. "The more, the merrier. Hey, turn on the radio and play me some tunes. I want to get warmed up."

An Andrews Sisters song came on, and Sherertz began dancing around the table as though he had a partner.

"Is he like that all the time?" asked Stern.

"Sometimes he's worse," said Drane.

"Hey, did you hear that Captain Valkenberg has canceled shore leave for tomorrow night, until after some dumb old sports awards?" asked Ensign Jones, who had his back to the room portal.

Meanwhile, Captain Valkenberg had just arrived and was standing in the shadows, watching their banter and the dancing unnoticed. He walked into the room. "Nice dancing, Sherertz."

They all turned to look, and someone yelled, "Attention."

Jones turned red with embarrassment.

"Good to see all you BB Wonders again," said Captain Valkenberg. "At ease, sailors. Where are the cards and the chips? Aren't you getting kind of a late start with the poker game?"

"A lot to catch up on, sir," said Garfield.

"Permission to speak freely?" said Booth.

"Speak away," said Valkenberg.

"What's up with Admiral Kidd canceling the football season?"

Valkenberg hesitated before answering.

It's clear he's deciding whether to tell us something, thought Sherertz. Something big is coming.

"CINCPAC just got back from Washington, and something's brewing," Valkenberg finally said. "Too much chatter being intercepted from Japanese broadcasts. The feeling is the Japs are trying to distract us from their real plans. Most of the speculation is that the Japs are going to attack Malaysia or the Philippines. We'll keep everyone on a short leash so we're ready for anything. Make sure your men are alert." Valkenberg stopped talking and looked around.

Sherertz's eyes followed his. While the men all looked a little wide-eyed, they also looked resolved. He was glad Valkenberg had decided to prepare them for what was to come.

"What about Pearl? Could torpedo planes or dive bombers take out our battleships?" asked Booth.

"The consensus is that it's too far from Japan and too shallow for a torpedo plane attack to succeed, but we're still sending out scouting planes daily just in case," said Valkenberg. "The dive bombers are something else. If the Japanese carriers get this far, the planes on our carriers and Ford Island should give them all they can handle."

Everyone nodded, seemingly reassured.

"Now I need to go prepare some remarks for some dumb sports awards ceremony I have to preside over tomorrow night," said Valkenberg. He cut his eyes toward Ensign Jones, who was conveniently looking in another direction. "Good evening, gentlemen. Hope it's a great game." He walked out of the stateroom.

"What a great captain you guys have," said Mote.

Jones poked his head through the portal and looked up and down the corridor to make sure it was clear. "It's a good thing he's the skipper and not Admiral Kidd. That guy has a permanent poker stuck up his ass."

Everyone laughed.

"Enough of the serious stuff," said Stern. "We are way behind. Who's dealing?"

"I am," said Garfield. "Five-card stud. One-eyed jacks are wild."

"Anybody seen the movie they're showing tonight on the *Nevada*?" asked Drane. "It's *One Foot in Heaven* with Fredric March and Gene Lockhart."

"Nah, it's about some minister," said Taussig. "Sounds kind of boring to me." He grinned at Rev Drane. "Now tomorrow night there's a good flick showing on the *Arizona*: *Dr. Jekyll and Mr. Hyde*. Since I've got deck duty Sunday morning, I'm thinking about going. Any takers?"

"I wish they had more John Ford movies," said Uhrenholdt. "I loved *Stagecoach* with John Wayne. Since we gotta stay here for the awards, and they're not showing the movie until after that, I'll go with you."

"I thought I might check out the Battle of the Bands at the Rec Center," said Jones. "Anyone interested?"

"After my stint at the academy, I'm kind of partial to bands myself," said Taussig. "If I go with Sherertz and company, I'll start off deck duty with a hangover, so come by JO Country and pick me up when you're ready to go ashore."

"We'll miss you, Taussig," said Sherertz. "Be careful if you get near the golf course. You might get beaned by one of our balls."

"Say, what's all this crap with having to lock up your

duffels when you go ashore?" asked Garfield.

"There have been a bunch of burglaries since some new sea dogs arrived," said Mote. "Until they catch the thief, we have to lock up."

And so it went into the night.

Shore Leave

Ensigns Sherertz, Drane, Mote, and Davis were riding the launch to the officers' dock. On the way, they watched the USS *Vestal*, a battleship tender, move into place alongside the USS *Arizona*, where the USS *Nevada* had been the night before.

"Just gas her up, check the oil, and clean the windows please," said Mote, as he watched the *Arizona* begin the refueling process. "Can you believe the *Arizona* is going stateside for Christmas?" said Mote.

"Mote, I told you not to buy those racy magazines," said Davis. "Otherwise we would be going with them." They all laughed. "Say Mote, do you loan out those magazines, or just keep them for yourself?" prompting more laughter.

As they neared the Navy yard, they passed the USS *Ramapo*. On its deck were two Patrol Torpedo boats, PT boats for short, about eighty feet long with men on deck busily working away.

"What do you use those for, waterskiing?" yelled Sherertz to the men on their decks, who promptly responded with a Bronx salute.

The *Nevada* ensigns all laughed.

As they got off at the officers' landing, Sherertz said to Drane, "Are you sure you don't want to come swimming or golfing or dancing with us, Rev? I hear the waves are breaking nicely over on Waikiki."

"Do I have to keep reminding you guys that I just got married?" asked Drane.

"I guess we know who wears the pants in your family," said Davis. "Now where are you going, Sherertz?"

"To see if there's anything interesting going on down at the hospital. Lieutenant Samuels told me last night I could round with him this morning. This time I even remembered my white scrubs. Don't worry, Davis, I'll be there in time to tee off."

~~~

Saturday Night, 1900 hours

Sherertz and his fellow ensigns walked into the front entrance of the Royal Hawaiian Hotel, just a few hundred yards down the beach from the Hotel Moana, where they were staying. The Royal Hawaiian was a favorite spot for service men on leave and it frequently held dances specifically for them. Now they were ready for the evening's festivities. The Royal Hawaiian, also called the Pink Palace, was built in 1927 adjacent to Waikiki Beach in a sparsely populated area of Oahu. Influenced by Spanish architecture, it had arches everywhere, delicate steel-grate balconies for each room, and was covered by pink stucco. Johnny flipped the bellhop a quarter to park the car, and they all went inside and headed down to the main dining room.

They walked inside between beautiful pillars supporting the massive roof. Just beyond they could see large glass chandeliers. Windows were plentiful around the perimeter wall, giving a beautiful view of Waikiki Beach during the daytime. The maître d' seated them at a table with a group of women who had already arrived and then handed out menus.

"Sherertz, what is all this crap?" asked Mote. "Paupiette *de Mahimahi Boieldieu*, butter palm leaves, Cream Doria in Jelly Romanoff. Do we eat this stuff or rub it on our dates?" He leered at their lady friends, who pretended to be insulted.

"You are such a country bumpkin, Mote," said Sherertz. "Permit me to order for all of us."

~~~

After eating, Sherertz and his group went to the Surf Room, where there was a live band playing, amid patriotic displays of the US flag. Sherertz and his date hit the dance floor, and the rest of their group followed. Most of the men there were Navy, but there were also some Marines and Army men.

The band, a particularly good one, finished the song they had been playing and launched into "Boogie Woogie Bugle Boy." Sherertz and his partner started dancing the jitterbug, Jack's favorite dance. A circle formed around them, with nearly everyone in the room watching. At the close of the song, they applauded and cheered for the two best dancers in the room.

Then the band played another song, and everyone started dancing again.

Jack looked around as he danced. He thought he noted a certain air of desperation in the frenzied dancing, as though everyone knew something was about to happen.

Many songs later the six couples headed back to their table. As they got there, they met up with Harold Christopher and Andy Uhrenholdt, both with dates.

"Hal, I didn't realize you were here," said Sherertz. "I thought you and Andy were going to the movie."

"Turns out the Jekyll and Hyde film was terrible, so I called up Helen and Uhrenholdt got a date, and here we are," said Christopher. "It's a good thing the carriers weren't here, or we wouldn't have been able to find parking."

"Well, you definitely came to the right place," said Mote. "After golf we went back to the Moana and showered and then went down to the Banyan Courtyard. Its 5PM and here we are again on beautiful Waikiki Beach with Diamond Head in the distance and surfers on the waves. This is Webley Edwards, Calling from Hawaii. Harry Owens and I will be bringing you the best music from Hawaii."

"Well, Mote, if we're stationed out her long enough, you

may get that Web Edwards imitation down," said Sherertz with a chuckle. "But I think you're going to have to work on the background music and conch shell horns, if you want it to really sound good."

"I'm just not into that Hula stuff, and what's up with the Willamette football team?" said Mote.

"They were here for the Shriner's Bowl," said Sherertz. "Don't you read any of the announcements?"

"You were wearing us out, Jack," said Helen, Ensign Christopher's girlfriend. "Where did you learn to dance like that?"

"In high school, my friends and I started a dance club called the Pinchee Winchees. Every Saturday night, we played music and tried out the latest dances. In college at Georgia Tech, there was a party almost every weekend, and I kept dancing. I almost flunked out of my first year at college. Who's ready to go again?"

"Jack, we're all exhausted and need a break," said Christopher.

"Harold, I think we should go practice some more so we'll be ready for our big day," said Helen.

Christopher looked nervous, then excited. Finally he blurted out, "Helen and I are getting married. We haven't told our families yet, but we had to tell *someone*."

Helen held up her hand to show them her engagement ring. The other girls crowded around her to look at it more closely, and Sherertz and Mote came over to congratulate

Christopher. Christopher reached out and shook Sherertz's hand, showing a ring on his own hand in the process.

"Nice ring, Hal," said Sherertz. "What does DTHS stand for?"

"I would take it off so you could look at it, but it's too tight. DTHS stands for Dwight Township High School. It's my high school ring. Pretty neat, huh?"

"Its swell," said Sherertz. "I like the Indian."

"Just the same, Helen, I need a blow," said Christopher.

"Before you go, I want to propose a toast," said Sherertz, picking up his glass. "To Helen and Hal. May you have a long life filled with happiness."

The others applauded.

"Hal, while you take your cigarette break, I need to make a phone call," said Sherertz, setting his glass down. "Y'all rest up." He walked away and returned a few minutes later.

"What was all that about?" asked Mote.

"Tomorrow, as you know, is the big boxing match between our men on the *Nevada* and the *Arizona*. Since it will be on the deck of our ship and Taussig has deck duty, I have the honor of being master of ceremonies."

"Who looks good, Sherertz?" asked Christopher.

"The last boxing match with the *West Virginia*, I lost my shirt betting against Dorie Miller," said Mote. "I need some cash for my big date next weekend. Come on, fess up,

Sherertz."

"Well, you didn't hear this from me, but . . ." Ensign Sherertz leaned over and whispered to Christopher and Mote. They laughed and nodded knowingly and smacked Sherertz on the back.

"This could be very profitable, after all," said Christopher.

Sherertz stood up and bowed with a flourish. "Ladies and gentlemen, I bid you good night."

"What's the hurry?" asked Mote. "It's only midnight, and the dance floor's heating up."

"Unlike the rest of you party animals, I have to get up at six thirty to get things organized, so I'll leave the dance floor in your capable hands," said Sherertz.

"We'll be up later, Jack," said Mote. "We promise not to slam the door too loud when we come in."

~~~

December 7, 1941, 0630 hours

It was morning, and the sun was streaming in the open window of the room at the Hotel Moana shared by Sherertz and his friends. The phone rang. Mote answered.

"If you aren't a beautiful woman, I don't want to hear it," said Mote.

"This is Chief Warrant Officer MacDougal at the officers' dock with Sherertz's zero six thirty wake-up call, as requested."

"Jack, it's your damn wake-up call."

"Thanks, Johnny."

Ensign Sherertz showered, shaved, dressed, and put things back in the small overnight bag he had brought with him. When he came out of the bathroom, the rest of his ensign buddies were getting dressed—although still not awake.

"Listen, Sherertz, we decided to share a cab back with you," said Mote.

"All right, here's the deal," said Sherertz. "I'm going down and calling a cab. If you're ready by the time it comes, you can join me."

~~~

In front of the hotel, Sherertz and three other ensigns piled into the cab. Sherertz sat up front and started talking with the cab driver. The rest got in the back and immediately sacked out.

They headed from the hotel straight out to Kalakaua Road and turned left. Both sides of the road were lined with palm trees.

It's really quite beautiful, if you're awake enough to appreciate it, thought Jack as he glanced at the sleepers in the backseat.

Quickly they left the trees behind, and Hickam Field became visible to their right across bare ground. Things were quiet. All the planes were parked in rows, with a few men working on the engines. They wound around Hickam

Field until they reached the entrance to the naval base. Seeing the sleeping sailors, the guard at the gate motioned them on through, and the cabby slowly drove past the Marine barracks with the drill field in front.

Through the open windows, Sherertz could hear marching commands: "Your left . . . your left . . . your left, right, left."

As he listened to the singsong cadence of the drill sergeant, he rubbed his nose. It was still sore from his collision with Mutt's drill rifle months previously. The cab driver swung around and pulled up in front of the Officers' Club. With much moaning and groaning, all the ensigns got out of the cab and stood up, blinking against the light. Sherertz paid the driver, and they started walking around the side of the Officers' Club toward the officers' dock.

Figure 16. Royal Hawaiian Hotel (left), Moana Hotel, pre World War II. University of Hawai'i at Manoa Library. Image can be found at www.airandspace.si.edu/explore-and-learn/multimedia/detail.cfm?id=7067 and at www.AHeroAmongMillions.com.

Figure 17. Route (dotted line) taken by Ensign Sherertz and fellow ensigns from Royal Hawaiian Hotel (above left), around Hickam Field, around the Marine barracks to Officers' Club (bottom). 1940 Photo, /Hawaii Department of Transportation, Airport Division. Image can be found at www.aviation.hawaii.gov/airfields-airports/oahu/hickam-fieldair-force-base/ hickam-field-photos-1937-1941/ and at www.AHeroAmongMillions.com.

Pearl Harbor Attack

December 7, 1941, 0730 hours

Jack Sherertz and his fellow Nevada ensigns arrived behind the Officers' Club. They walked slowly toward the officers' dock along a well-worn path. Straight ahead across the water was the submarine base with the Commander in Chief of the Pacific's (CINCPAC) office on the top floor of the main building. Just to the right was Merry Point, and behind that could be seen the large tank farm storing fuel oil. To their left was the 1010 dock with multiple ships in dry dock for repairs. The closest ship was the destroyer USS *Bagley*. Sailors were going to and from the ships all along the dock, and trucks and cranes were visible loading supplies. Farther left and across the channel was Battleship Row, with Ford Island right behind. Directly behind where they stood and adjacent to the Officers' Club were two flagpoles. An honor guard had just finished raising the Hawaiian flag. It rippled gently in the breeze against a cloudy sky. The honor guard had the US flag in hand, ready to raise it at 0800, as was customary each morning throughout Pearl Harbor.

Sherertz saw Chief MacDougal standing on the dock all by himself and walked over to him. "Chief, if this day gets any slower, we're going to have to bring you a chair, an umbrella, and a cold drink. Where the hell are all the launches? Is someone having a boat sale?" Sherertz laughed.

"They're all taking sailors back to their ships—sailors who got here on time," quipped MacDougal.

"We need to get back to the *Nevada* by 0800," said Ensign Christopher. "What's it going to cost us?"

"It's not my fault all you ninety-day wonders missed your ride." MacDougal threw back his head and laughed, his ample belly jiggling.

"Where's that fishing boat of yours?" asked Ensign Uhrenholdt. "I know you keep it around here under the dock. I've seen you take our skipper out on it. We'll send it back with our *Nevada* officers taking Sunday leave."

"If you think I'll loan my pride and joy to a bunch of green ensigns, you're all crazy," said MacDougal.

"What if we sweeten the deal with a fifth of scotch?" asked Christopher.

"Good luck getting MacDougal to crack," said Sherertz, shaking his head as he and Ensign Mote walked away from the negotiations. "Check out those planes over Pearl."

High above Pearl Harbor, Jack saw several isolated planes circling and a line of planes coming in from the south. *Very strange*, he thought. Aloud, he asked, "Do you think the carriers are back?"

"Do you see any carriers?" retorted Mote.

"You're such a wiseass, Mote," said Sherertz. "They could be just outside the harbor."

"Right . . ." Mote laughed as Christopher and Uhrenholdt joined them.

"Maybe they're PBYs (Patrol Bomber amphibious aircraft made by Consolidated Aircraft) taking off late for their reconnaissance runs," said Sherertz. "But their wings don't look right." The hairs on Jack's neck were beginning to stand up.

Unnoticed by the group, Rear Adm. Patrick N. L. Bellinger, Commander of the Navy's Pacific air wing, had come up behind them, accompanied by two lieutenants, a lieutenant commander, and a captain. They were all dressed in white uniforms. The admiral had two stars on his collar and two stars on his gold braid epaulettes.

"No PBYs will be taking off late on my watch," said Bellinger, as he looked up at the planes, concern on his face.

The Nevada ensigns turned around, saw his two stars, and snapped to attention. "Aye, aye, Admiral, sir."

"At ease, men," said Bellinger. "You need a ride?"

"We need to get to the *Nevada*," said Sherertz.

"My staff will be taking Captain Bode back to his ship, the *Oklahoma,* and then we're going to Ford Island. That will get you halfway there. Hop in."

One of the two lieutenants got on the launch first and went to the helm. He checked the gauges and started the engine. The other lieutenant remained on the gangway.

Then Bellinger got on and went to the front on the port side, followed by Bode, who sat opposite him on the starboard side, and finally the lieutenant Commander sat near Bellinger. The four *Nevada* ensigns scrambled on last and walked to the back of the boat. The floorboards creaked with each step they took. When all were seated, the lieutenant on the gangway cast off the mooring lines and came aboard, moving to the rear of the boat, where he took hold of the rudder. The piloting lieutenant increased the throttle slowly, while the aft lieutenant steered the launch away from the dock.

Mote leaned over to Sherertz and spoke in his ear. "Is that guy Bode, Charley Stern's new skipper?"

Figure 18. Pearl Harbor ship positions, December 7, 1941, 0755. Drawn by Robert M. Berish for The Rising Sun in the Pacific. Numbers over water indicate depth in feet at low tide. Original is kept at US Naval Historical Center. Image can be found at www.navsource.org/Naval/helpers/pearlmap.jpg or at www.AHeroAmongMillions.com.

Sherertz nodded. "It's a good thing neither you nor I have watch at zero eight hundred like Taussig, because we may not make it." As if to rub it in, they heard the band from the *Nevada* begin warming up to play "The Star-Spangled Banner."

"That plane looks like it's diving on Ford Island," said Bode to Bellinger. "What are those planes up to?"

Figure 19. December 7, 1941. Japanese planes over Pearl Harbor at start of attack. Japanese aerial photo. View is looking east across Ford Island with Battleship Row just behind it. USS West Virginia had just been struck by a torpedo producing the visible water plume. This image may be found at history.navy.mil, Photo # NH 50930 or at www.AHeroAmongMillions.com.

"Damn, I'm sure as hell going to find out!" Bellinger retorted, as he stood and gripped the boat rail.

Boom, boom, boom came from behind them, from the direction of Hickam Field. They all turned to look. Planes were visible above Hickam Field. Two were in a steep dive.

"That sounds like bombs," said Sherertz.

"No shit, Sherertz," said Mote.

This is it. It's finally happening, Sherertz thought, as he shuddered down deep inside.

"Fuck, if they hit the tank farm, we're all going to burn,"

yelled Christopher.

Boom, boom . . . boom, boom, boom came from the direction of Ford Island, and they turned back around. A plane in a steep dive disappeared behind the battleships, and more planes were in a line directly above.

Suddenly they heard the USS *California* across the channel sound the general quarters alarm, followed almost simultaneously by the USS *Cassin* in 1010 dry dock, and then ships all over the harbor chimed in.

"Battle stations, battle stations," blared the PA system of a ship off to their port side. "This is not a drill. This is not a drill."

"What the hell!" yelled Mote in a high-pitched voice. He stood up, holding on to the rail. Everyone else stood and looked toward Ford Island.

From Merry Point behind them, five airplanes buzzed over at an altitude of thirty to forty feet, a red ball visible on their fuselage and the bottom of their wings, made even more iconic by the rising sun behind them, visible through a break in the clouds.

"Jap planes. Damnation!" shouted Bellinger. "Full speed!"

Jack heard the boat throttle up and felt it jump forward, but he felt as if it were all a dream. He had trouble convincing his brain that this was real. Then the first three Jap planes dropped their torpedoes in front of them, heading in a direct line toward the USS *Oklahoma.* That snapped him back to reality. The fourth plane dropped its torpedo, heading toward the USS *West Virginia*. The turbulence lines from the torpedo propellers left no doubt that impact was imminent.

Figure 20. Japanese torpedos coming from Merry's Point about to impact USS *Oklahoma*. Fires at Hickam field in background. Japanese aerial photo, US Navy National Archives, history.navy.mil Photo # 50931.

"Aiiee!" yelled Captain Bode as he violently kicked the side of the boat and clutched the rail with white knuckles, an anguished look on his face. "My God, they're trying to sink my ship! Damn them!"

To their left, Jack saw men on the USS *Bagley* sitting in dry dock open fire with their forward machine guns. They missed the fourth plane but hit the fifth. Guns were opening fire all over the harbor.

"Fuck you, Japan!" yelled a *Bagley* sailor as he fired a rifle at the plane their machine guns had just hit.

The Jap plane's left wing disintegrated, and it plunged into the water, right in front of the admiral's launch.

"Hard to port!" yelled the admiral.

The lieutenant swerved left and just missed it. This maneuver brought them near the docked USS *Ramapo*, where both PT boats on its deck were firing their .50-cal machine guns at the planes above. The shell casings bounced off the deck and rained down from above.

"Give 'em hell!" yelled Sherertz, as he felt a sudden surge of adrenaline-fueled anger.

Back on course toward the *Oklahoma*, they saw three torpedoes strike it in rapid succession. The explosions pushed the port side of the massive ship up in the air, along with towering plumes of water. It came back down, beyond horizontal, listing toward the port side, as three gaping holes took on water. Jack saw the look of horror on Bode's face get progressively worse.

As the *Oklahoma* listed more and more to port side, the mooring lines tethering the *Oklahoma* to the *Maryland* stretched tight like piano wire, initially slowing the process. *Pow, pow, pow.* The mooring lines snapped sequentially like loud gunshots, strong enough to knock men over on the *Maryland* as the ropes flew back over their deck. The *Oklahoma* now rolled unimpeded toward the channel. Men on deck scrambled up toward the starboard side, climbing over the edge if they could. Some didn't make it and slid off into the water, where they franticly swam away from the boat to avoid it capsizing on top of them. Jack felt helpless.

"Oh no, fuck me," said Captain Bode. His legs almost buckled as he watched his ship sink before his eyes, unable to do anything.

Another plane went by them on their port side and dropped a torpedo aimed at the USS *California*. It veered up and to the left as it did. The USS *Avocet*, just in front of the *California*'s bow, opened up with antiaircraft fire. Smoke and fire erupted from the plane's fuselage, and it rolled down and to the left, crashing into Hospital Point with a loud explosion.

"Attaboy, *Avocet*," yelled Mote.

Jack wondered if this was the beginning of an effective counterattack. Then he saw something that made him think otherwise. From the right, a line of dive-bombers came down from the rear of Battleship Row, dropping their bombs as they flew over.

Boom, boom, boom.

Bombs struck the Vestal first and then the bow of the *West Virginia* on their starboard side, and finally the *California* on its port side. All three ships erupted in flames. The flames were so intense on the West Virginia that men were jumping off into the water all along the port side. Even though they were a few hundred feet away, Jack could feel the heat of each explosion.

"*Whoooooooomp,*" came from their starboard side, loud enough to hurt their ears.

Figure 21. USS *Arizona* blowing up. At far right is the mast of the USS *Oklahoma* heeled over to port as she capsizes. Photo taken from North of Ford Island. US Navy National Archives, history.navy.mil Photo # 80-G-6683.

Sherertz turned to look in time to see a cloud of flames and smoke rising five hundred feet above the USS *Arizona*. He could make no sense out of this. Before he could think about it further, the shock wave hit, knocking him and the other men on the launch backward to the deck. The lieutenant at the stern was nearly blown out of the launch, and Jack hit his head as he fell back, momentarily stunning him. When he got up to his hands and knees and shook his head back to full consciousness, he saw that Mote was still down. He helped him get up, and then both of them turned and looked again toward the *Arizona*. It was already sinking. Suddenly shrapnel and other *Arizona* debris rained down, much of it on fire. Jack's ears were still ringing from the explosion.

A near miss of their launch produced a whooshing sound, accompanied by a geyser of water more than two hundred feet in the air, raining back down on them.

More planes became visible diving on Ford Island, followed by explosions and towers of smoke.

"Abandon ship!" screamed a voice through the *West Virginia* PA system.

The *Oklahoma* was now listing more than 45 degrees to port, rolling faster all the time. Plates and trays from a galley below deck came flying out of a stairwell, and ammo for the five-inch guns fell out of the turrets, both mixing together in the air, clattering as they fell into the water.

Small boats were visible, heading to and from the battleships. One was strafed by a Jap Zero. Jack saw the machine-gun rounds tear through the men on board, exiting their bodies with bursts of red spray. The boat stopped moving, as water spurted up through the holes in the craft like small fountains. Jack felt sick and nearly threw up, but held it down.

Left of the *Oklahoma* was the USS *Neosho*, so far untouched.

As they closed in on the *Oklahoma*, it finished capsizing. Some of the swimming men were trapped underwater as it did, their screams for help silenced. Jack felt another surge of anger and frustration at not being able to help these men. The *Oklahoma* would have turned completely over, but its main tower hit the bottom with a crunch, sending a shudder through the whole ship and stopping the rotation.

On the bottom of the ship, which now faced the sky, stood the lucky few men who were able to escape the deck. Some waited for the small boats heading in their direction. Some just dove in and started swimming toward the *Maryland*. Bode writhed in the agony of the moment, the agony of his dying ship. He finally had to turn away, unable to deal with the terrible scene he had just watched unfold.

"Admiral, drop me off at the *Maryland*," said Bode in a strained, resigned voice. "I can't do anything more here."

Figure 22. USS *Oklahoma* capsized with USS *Maryland* to left and the burning USS *West Virginia* just behind. US Navy National Archives, history.navy.mil, Photo # 80-G-19949.

The launch pulled over against the *Maryland*, and the men on deck dropped a rope down and pulled Bode up. He immediately ran to a ladder and climbed up to the bridge. The admiral's launch pulled away from the *Maryland*, hugging the port side of the *Neosho* to avoid the intense heat coming from the burning *California* and surrounding oil fires, and pulled up to the Ford Island dock.

"Men, I was going to keep my launch here, but I think the *Nevada* needs you," said Bellinger. "Lieutenant, get them to the *Nevada* as fast as you can. Pick up survivors along the way. Good luck."

"Now you're talking," said Jack.

The admiral and the lieutenant commander quickly exited the boat and ran off the dock. They disappeared into the smoke coming from Battleship Row, mixing with the smoke from Ford Island.

The lieutenant at the helm of the admiral's launch backed away from the dock and took off as fast as the debris in the water would allow, steering back past the *Neosho* on a path to go to the right of the *Oklahoma*, directly into the channel. All the *Nevada* ensigns were standing, looking for planes.

"No, no, hard to port!" yelled Christopher, attempting to be heard over the now unimaginable din created by all the explosions and gunfire. "Jap planes are strafing the channel." The lieutenant turned the wheel hard to port, and two parallel lines of machine-gun fire hit the water to their starboard side.

Jack suddenly found a pistol in his hand, firing at the passing plane.

Boom, boom.

Two more torpedoes slammed into the *California* off to their right, sending geysers of water into the air, followed by the sound of an avalanche of water pouring into two dark holes in the ship's hull. It began to list to port.

The launch pulled in between the capsized *Oklahoma* and the *Maryland*. Men on the *Maryland* deck were throwing down ropes to pull up survivors from the *Oklahoma*. The roar of the three-inch and five-inch antiaircraft guns on the *Maryland* was deafening. As the launch went between the two ships, two men waved frantically and then slid down the hull of the capsized *Oklahoma* into the water. It was soon apparent they couldn't swim, and the launch had to go get them. Sherertz and Mote pulled them in. Other survivors were visible in the water, but nearby boats were heading in their direction, so they kept going. They could see a few men up on the *Oklahoma* hull, trying to make an opening to get some men out.

Clank, ... clank, ... clank, a heavy, rhythmic sound came from inside the hull, as though someone was swinging a heavy wrench, trying to signal or knock a hole in the hull.

The lieutenant piloting the boat headed briefly toward the gap between the *West Virginia* and the *Tennessee*, but he couldn't see a thing because of the smoke, and the heat from the burning oil between the two ships was unbearable. The only option was to steer between the sunken, capsized *Oklahoma* and the burning *West Virginia* and turn left toward the *Nevada*.

Figure 23. USS *Oklahoma* capsized (left), USS *Maryland* just behind it, USS *West Virginia* burning furiously on the right. US Navy National Archives, history.navy.mil, Photo # 80-G-33035.

As they went by the port side of the *West Virginia*, a wall of smoke, blown with the prevailing wind toward the southwest in line with Battleship Row, nearly obscured the port side and rose up behind the battleships. Each time the wind shifted their way, they were coated by the smoke's oiliness and could not avoid breathing the acrid fumes. Fits of coughing followed, bringing up foul black stuff from deep inside their lungs.

Figure 24. Launch moving toward the burning USS *West Virginia*. US Navy National Archives, history.navy.mil, Photo # 80-G-19930.

Approaching the end of the burning *West Virginia*, they saw two men struggling to swim through a dense layer of black oil on the water's surface. Christopher and Mote pulled them into the boat, with oil plopping off their bodies like thick, black pudding that congealed on the boat floor. They both crouched on their hands and knees, coughing crazily, trying to rid themselves of inhaled oil and water.

Suddenly from behind the stern of the *West Virginia*, a boatful of men burst out across the launch's bow, heading toward the channel.

Boom!

A bomb struck the boat, and it disappeared into the air. Debris flew in all directions. The concussion was deafening, and the shock wave almost knocked the men out of the launch. A plume of water went up with the explosion and came down all over them.

"Oh my God!" yelled Mote.

There on the floor of the launch was an arm with a tattoo and blood pouring out of it.

"Aiiieee! It's . . . it's . . . one of our bosun's mates," said a sailor they had just pulled out of the water. He began to shake uncontrollably and retch. Everyone who saw it flinched.

Passing the stern of the *West Virginia*, they saw a man going from the West Virginia to the *Tennessee* via a rope strung between them, thus escaping the inferno on the *West Virginia* deck. Some men weren't so lucky and came running off the *West Virginia* deck with their clothes on fire, jumping into the burning oil below. Out of the burning oil, Sherertz and Mote pulled a few badly burned survivors, hair singed off, clothing seared to their skin. Some of the skin came off in their hands as they pulled them out. It crunched. Mote threw up. Some of the burn victims were in shock and said nothing. Others screamed loud enough to be heard over the guns firing and the explosions. The smell of burning flesh caused Jack to shudder with revulsion.

Dead bodies and body parts were everywhere. Some of the body parts still had clothing on them. A disembodied floating head had a sailor's cap, fused to it by the flames.

On their way past the *Vestal*, they saw a man ahead waving frantically to them and yelling for help. They headed toward him, only to have to turn hard to port to avoid a Japanese strafing run coming from directly behind him. When the plane was gone, they saw the man still there holding up his arm to be pulled out, but he was no longer yelling. The water around him was intensely red, and the man's eyes seemed lifeless, his face white. Sherertz and Mote lifted him out of the water. Both of his legs were gone, and blood poured from the stumps, dripping down the shreds of still-attached flesh. Reflexively, they both let go and dropped him on the floor of the boat. He shuddered once and died. The two men closest to him in the launch moved to get farther away. Jack felt a growing sense of detachment as his mind struggled to cope with what he was seeing. They kept moving, looking for survivors. No more were found.

They passed the stern of the *Vestal*, and there was the *Nevada*, backing up from the *Arizona* and the *Vestal*. In front of their launch, Jack saw bodies facedown in the water. They turned a few over and found they were sailors from the *Nevada*, blown off the deck by the explosion that sank the *Arizona*.

Boom. A torpedo struck the port bow of the *Nevada*. Debris from the explosion flew in their direction, requiring them to duck. In slow motion, the *Nevada* rose up into the air, showing a gaping wound in its side, and then settled back down as water poured in.

"Lieutenant, we need to get there now!" yelled Jack, feeling the urgency intensify as he stared at the looming hull of the *Nevada*. Finally, they reached the starboard-docking platform. The *Nevada* ensigns leaped onto it and bounded up the gangway.

USS Nevada

Figure 25. USS *Nevada* near 1010 dock heading down channel toward ocean (to the right). Ensign Sherertz and fellow ensigns on board. Taken from Ford Island. US Navy National Archives, history.navy.mil, Photo # 97397.

Reaching the bow of the *Nevada*, Sherertz, Mote, and Christopher could barely see anything because of thick, black smoke. Suddenly the wind blew the smoke away so they could see.

"This is Commander Thomas," came a voice over the PA system. "Sherertz, you and your fellow ensigns organize the firefighting."

Ensign Sherertz ran back along the port side to get an ax

to clear some deck wood that was burning and obstructing foot traffic. Mote grabbed a seaman and the two of them found a hose and began trying to put out fires. Christopher climbed up to the five-inch antiaircraft director to see where he could help.

As Jack ran toward the bow of the *Nevada*, he saw downed seamen everywhere. He checked each one. Some were dead. Some had obvious mortal wounds — one lay there seizing with a head injury, his brain pouring out on the deck. Another had a piece of shrapnel in his chest, blood spurting like a geyser around it. Another was holding his intestines in his hands. *They are killing us*, his brain screamed. Sherertz fought the urge to throw up and the equally powerful urge to run away. Miraculously, some men had no visible injury, but were lying there stunned by the shock wave. He saw a seaman coming around from the port side.

"Yo, Seaman, get this man over to the doc," said Sherertz.

Advancing to the bow edge, he looked forward. Men were pouring off the stern of the *Arizona*, trying to escape the wall of flame advancing from the bow of their boat. He couldn't help them, so he turned around. He found a seaman struggling with a hose and helped him put out a deck fire until another seaman came along. When it was clear that all available hoses were in use, he went back to assisting the corpsmen identifying men who might still be saved.

Suddenly Chief Boatswain Hill appeared on the foredeck, soaking wet, having just cast off the *Nevada*. There followed the unmistakable shudder of the *Nevada's* screws changing from reverse to forward while the boat turned to port side. As it did, the *Nevada* moved closer to the flaming *Arizona* and unbearable heat. The incoming tide caused the *Nevada* to pick up speed slowly.

"Give 'em hell, *Nevada*," yelled a seaman in the water just aft of the *Arizona*. Whistles and cheers followed from everyone who could see her progress. Sherertz felt their emotional response echoing his churnings. *Finally something was happening.*

A gust of wind briefly cleared the deck of smoke, and Sherertz saw Ensign Taussig being lowered down from the port side five-inch antiaircraft gun director on a stretcher. He ran over to help lower him to the deck. Taussig's left leg was sticking out at an impossible angle, and it was hemorrhaging in spite of a corpsman holding a pressure dressing on the wound. His lower body was covered with blood. His face was pale.

"Taussig, I can't leave you alone for a minute," said Sherertz.

"This wouldn't have happened if you weren't out partying all night," said Taussig.

"Corpsman, has the doc seen him yet?" Sherertz asked.

"He's been a little busy, sir."

"Listen, I'll hold pressure on the wound while you go

get the doc."

"I don't know if I should leave him, sir."

"Go now!" commanded Sherertz.

"Yes, sir. I'll be right back."

Sherertz pushed on the dressing over Taussig's leg, and blood spurted out, accompanied by loud groans from Taussig.

"This is not good, Taussig."

"You're telling me."

"So what the hell's going on here, Taussig?"

"I think the Japs were using me for target practice."

From behind him Sherertz heard a voice say, "Can I swap places with you, Ensign."

Sherertz turned to see the doc. "He's all yours, Doc," he said. "Take good care of him. He owes me money. Besides, now that we're getting up to speed, I've got a war to fight."

Unfortunately, the *Nevada's* movement down the channel had not gone unnoticed.

"Sherertz, you better get ready," said Taussig. "I think the action is about to really heat up." He pointed skyward.

A whole line of Jap planes was visible, heading straight for the *Nevada*. And then the bombs came, gleaming in the sky. First there were two near misses that sent columns of water hundreds of feet into the air, raining back down on the deck. Then bombs began to connect. In rapid succession, three bombs struck the USS *Nevada*, one right near the forward fourteen-inch guns. Each explosion rocked the ship like an earthquake, throwing men to the deck, and burning debris flew in all directions. Sherertz leaned over Taussig to shield him.

The *Nevada* began to list to starboard and take on water. Suddenly, as they went between the dredge and the USS *Shaw* in floating dry dock, there was an enormous explosion from the *Shaw*. More burning debris showered down on the *Nevada*, and more fires were ignited on the already burning *Nevada* deck.

Figure 26. USS *Nevada* continuing down channel a few minutes later. Photographed from Ford Island looking across the USS *Avocet* and the dredge line to the *Nevada*. Bow of *Nevada* is now on fire, having just been bombed. US Navy National Archives, history.navy.mil, Photo # NH 97396.

The PA system blared. "All hands not involved with

gun positions or ammunition, get out and fight the fires."

Unexpectedly, the *Nevada* turned hard to port, toward Hospital Point. CINCPAC had ordered it beached before it could sink and block the channel. The bow ran aground, jolting the entire ship. To hold it there, they had to drop anchor. Chief Hill sprinted to drop the bow anchor and had started it down when another bomb struck the bow, killing him instantly. The incoming tide was now pushing the stern of the boat toward shore, out of the channel. Seeing this, the Jap planes moved away, looking for other targets.

Sherertz continued to make the rounds looking for survivors. Right in the midst of this extraordinary chaos, an overdue friend arrived.

"Heh, Sherertz, where do you need me?"

Sherertz turned around to find Ensign Drane. "Where the hell have you been?" said Sherertz.

"I had a brief stopover putting out fires on the *California*, but I'm here now."

"Is Marty okay?" asked Sherertz about Drane's wife.

"Just scared."

"Take command of that group trying to put out the fire below our fourteen-inch guns and move any bodies you find over there with the rest."

Rev Drane headed over to take charge of the seamen and came across a body a few feet from where a bomb had ruptured the deck. The man had a gaping hole in his back and was obviously dead. "You two seaman. Come move this body!" said Drane.

They rolled him over so he would be easier to carry. His nametag said Davis, underneath a set of pilot wings. Drane didn't see him because he had already moved over to the next group of seamen.

~~~

A tug carrying Capt. Francis W. Scanlon, the captain of the *Nevada*, arrived and pulled up between the *Nevada* and the channel near Hospital Point. It turned so that it was perpendicular to the shore and pushed its bow until it made contact with the *Nevada's* starboard side. After Scanlon boarded the Nevada, the tug slowly pushed the *Nevada* toward Hospital Point. The deck crew quickly dropped the portside anchors and turned on the winches, pulling the *Nevada* even closer to shore. Then they tied cables and rope to anything they could find on shore, trying to hold her there.

Scanlon finished watching the winching and turned and met Lt. Commander Thomas coming his way. Thomas saluted.

"Thomas, status report," said Scanlon.

"One torpedo hit the port bow at frame forty-two, flooding all compartments forward of frame sixty," said Lt. Commander Thomas. "At least six bombs hit: one damaged all officers' quarters forward of the wardroom. One hit frame twenty-five, destroying the wardroom and damaging the anchor vertical shafts and the gun barrels of Number

149

One turret. One hit frame twenty-seven just forward of Number One turret, damaging the forecastle. One hit the boat deck about frame eighty, badly damaging the galley. The last one hit just forward of the port antiaircraft director, burning out the sky control shack and the navigation bridge and destroying your office and quarters. Water is up to six feet below the forecastle deck. Forty are known dead. There are more than ten missing, and dozens of men need to go to the hospital."

"Damn, we're scuttled," said Scanlon. "Any good news?"

Thomas glanced up. "The Japs have stopped bombing us," he said with a weak grin.

"I would say that under the circumstances, you men did a hell of a job," said Scanlon. "What's next?"

They both felt a slight movement of the *Nevada*.

"Did you order that?" asked Scanlon.

"No," said Thomas. "I think the incoming tide is floating us off the bottom."

"Get those tugs back here! Now!" said Scanlon. "Move us over to Waipio Point. The bottom's more stable there."

Figure 27. USS Nevada beached on December 7, 1941, near Waipio Point. USS Hoga is alongside her port bow pumping water onto the deck fire. US Navy National Archives, history.navy.mil, Photo # 80-G-19940.

Attack Aftermath

Sherertz and the other officers watch Scanlon return after a quick tour of the boat.

"Deck Ensigns, get your men together and put these fires out. The rest of you man the guns and be ready for anything. CINCPAC thinks we'll be attacked again. And get the ammunition conveyors fixed so we don't have to carry shells and powder by hand as you did earlier this morning. It's too damn dangerous. Any questions?"

No one voiced any.

"Then get to it."

Everyone took off, double-time.

"Drane, why don't you take the midship fires, and I'll take the forward fires," said Jack.

"That works," said Drane. "Say, have you thought about an old-fashioned bucket brigade?"

"We've been a little busy dodging bullets and bombs," said Sherertz. "Maybe later."

Putting out the fires seemed nearly impossible. The deck was shattered in multiple places, making movement difficult and dangerous, especially with smoke covering the deck most of the time. Only a few hoses worked because of bomb damage.

Two of his men came running toward Sherertz with a man on a stretcher. "He just passed out, Sir," said one of the seamen. "What do we do?"

"Take him aft, out of the smoke," said Sherertz. "Tell the rest of the men to wet their t-shirts and tie them over their mouth and nose. Tell anyone who feels dizzy or short of breath to get aft quickly and stay there until they feel better."

Meanwhile, the *Nevada* crew had gathered the wounded on the port side. As launches were available, they were taken to the hospital.

Finally the cavalry arrived in the form of the seaplane tender USS *Avocet* and some tugs: the USS *Hoga* (*YT-146*), the USS *Osceola* (*YT-129*), *YT-152*, and *YT-153*. From all directions, seawater poured down on the fire, and after an hour and a half, some of the deck fires were out. The tugs then collectively moved the *Nevada* across the channel to Waipio Point, where they pushed her up on hard ground.

~~~

By 2030 that night, everyone on the *Nevada* was exhausted and on edge. Most of the fires were out except for some burning below deck, but explosions continued from all over the channel as fires set off various munitions. Drane and Sherertz had deployed their observers to their observation posts.

Suddenly a bow seaman yelled, "Planes coming from the south toward Pearl Harbor!"

Drane relayed the message to Captain Scanlon on the bridge.

Scanlon grabbed a pair of binoculars, confirmed the sighting, and called the Pearl Harbor control tower. "Are we expecting any planes?" he asked.

"From what direction?" said the voice over the radio.

"Due south," said Scanlon.

"Four planes are supposed to be coming from the *Enterprise*, but not from that direction," said the voice. "They might be bogies."

Before anything else could be said, gunfire erupted from all over Pearl Harbor. It looked like a fireworks display. There was so much tracer fire you could read a newspaper. Two planes were quickly shot down, and two turned around and headed back out to sea.

Up on the bridge, Captain Scanlon was still talking with the control tower. He put down the microphone.

"We just shot down two B-17s from the *Enterprise*," said Scanlon. He walked out of the bridge without saying another word.

~~~

Finally, well after midnight, the harbor was quiet. Sherertz and Drane had divided up their deck observers into two-hour watch stints, and the rest had found places to catch a few winks. Drane had taken the first watch with the deck observers, so Sherertz looked around for a place to land. He found a large coil of rope and curled up on it, desperately trying to block out the day's events enough to sleep. He failed.

~~~

The next morning, reveille sounded at 0530. Most were not asleep.

Scanlon addressed his officers on the aft deck at 0600. "I want all you deck ensigns to get your seamen to sweep the ship one last time for survivors. Then load up the bodies on the launches and head up to Aiea Landing. Remember, one seaman in each group carries a bucket for body parts or anything that can identify a seaman — dog tags, tattoos, et cetera. Bring them back to Ensign Sherertz to record. The rest of you come with me."

Sherertz and Drane walked over to one side. "So I'll handle the body search and you'll organize the watch," said Sherertz and Drane nodded.

"You four go with Ensign Drane," said Sherertz. "The rest of you stay here with me."

A group of smoke-stained men, many still in their underwear, lined up in front of Sherertz.

"Where's Fireman Harker?" asked Sherertz.

"He didn't make it, sir," said one of the seamen in a choked voice. "He got strafed while trying to put out a fire."

"No time to worry about him now," said Sherertz. "Groups of three, just like the drills. Now get moving."

156

They ran off and started sweeping the deck for survivors. Some went below to look in compartments that weren't flooded. Under Sherertz's direction, they piled bodies up on one side of the deck. Forty-plus bodies made a big pile, and a horrible stench.

Shortly, Drane returned and walked over to Sherertz. He asked to see the clipboard with the casualty list. "Damn, Sherertz. They took out most of one Marine platoon."

"They were all standing next to where one of the bombs hit," said Sherertz.

"They got Chief Hill," said Drane. "I really liked him."

"He was trying to lower the anchor after we ran aground, and a bomb got him," said Sherertz.

"Oh God, here's another one of our men," said Drane, pointing to a name and trying hard to keep his voice from quavering. "He told me he wanted to be a mechanic when he got out of the Navy."

"I don't even know how he died," said Sherertz in a somber voice. "Brace yourself, Rev. They got Curt Davis."

Drane froze in his tracks and stood there rigid, his face tight with emotion. For a moment, he didn't move. Then he slowly composed himself and turned toward Sherertz.

"He was standing next to our forward fourteen-inch guns when one of the bombs hit," said Sherertz. "He didn't have a chance."

Drane stared toward the bow briefly and then said, "Where are they taking them?"

"I heard Scanlon say something about a place called Punchbowl Cemetery."

Just then one of the seamen returned carrying his bucket. He held it out and, with a choked voice, reported. "I didn't know what to do with this, sir."

Ensign Sherertz looked into the bucket. There on the bottom was a ring with the letters DTHS on it. Sherertz's face blanched, and he shuddered.

"Are you okay?" asked the seaman.

"Leave the bucket here," said Sherertz. "Get . . . get back to your team and keep looking for survivors."

The seaman ran off. After he disappeared around the corner, Sherertz ran to the side of the boat and threw up. Drane picked up the bucket to look at what Sherertz had seen, and his legs started wobbling. He sat down and began to sob.

~~~

Ensign Sherertz and a group of seamen pulled away slowly from the USS *Nevada* gangway, their launch fully loaded with *Nevada* dead, and headed north through the soft rain from Waipio Point to Aiea Landing. The harbor was quiet, except for the low hum of other launches moving about. They continued slowly, trying to avoid the debris and floating bodies. Today was not their turn to pick up bodies.

When they arrived at Aiea Landing, Sherertz directed the men to unload their cargo. They placed the bodies in simple, hastily built, pine boxes, each with a metal tag that had a number on it. Taking care to see which numbered box each man was placed in, they added that number to a master list. They processed body parts separately, took fingerprints where feasible, and recorded tattoos. Then the coffins were put in the back of waiting trucks.

Sherertz and his men all got into another truck and headed out. Soon they arrived at Punchbowl Cemetery, where they quickly unloaded the coffins and took them to a hastily dug mass grave. A bulldozer waited off to the side, and on the other side a large group of Navy men of many ranks gathered. Sherertz walked around to the far side where there was more room. He turned and looked in the same direction as everyone else. From this direction, the mass grave almost looked like the shape of an anchor.

A short service followed, with a chaplain and a priest each saying a few words. Then some of the men took shovels, and others used their hands to toss dirt on top of the coffins below. Another group of men did the same, and then another. Sherertz took his turn. With tears in his eyes, he took one last look, turned, and walked away, followed by his men. As they left, another group of men finished covering up the coffins, each marked by a single wood stake. A number on the stake served as the only proof of the man's death at Pearl Harbor.

~~~

Sherertz and several other shipmates arrived at the Naval Hospital to visit Taussig Monday afternoon. None of them looked as if they wanted to be there. A harried seaman at the front desk gave them Taussig's room number, and they walked up the stairs and down the hall until they arrived at his ward. They walked in to find rows of injured seamen. Some made eye contact, some stared ahead with blank looks, others were unconscious or sleeping. Finally they spotted Taussig and walked over to him, trying not to look closely at the other men, afraid of what they might see.

"What's the good word, Taussig?" asked Sherertz. "When are they going to let you out of here?"

"No good words at the moment," said Taussig.

Without warning, he pulled back the covers, revealing that his left leg had been amputated above the knee.

Visibly horrified, everyone was speechless.

Taussig covered his leg back up. "How bad's our ship," he asked, breaking the silence.

"They shot it all to hell," said Drane. "It's beached over on Waipio Point. We're all being reassigned."

"Anyone get their orders yet?" Taussig asked.

"We'll hear next week," said Sherertz. "Meanwhile, we're cleaning up your mess."

Everyone laughed nervously. When Taussig moved and winced with pain, some of the men around the bed turned their head, unable to watch.

"It's about time for a beautiful nurse to come give me a

bath, so you ugly mugs need to get the hell out of here," said Taussig, clearly showing the ongoing pain in his face.

"We'll check back tomorrow," said Sherertz.

As they walked out of the room, Sherertz saw Lieutenant Samuels, who had been taking him on rounds.

"Sherertz, why don't you hang out with me today?" said his doctor friend. "We're really busy, and I could use the help."

"No, thank you, Samuels," said Sherertz with a strained look on his face. He turned and walked away without saying another word. *The thought of seeing all those wounded men made him feel sick, deep deep inside. He couldn't wait to leave.*

Another physician walking with Samuels asked, "What's his problem, Samuels? Doesn't he know you could get in trouble for taking him on rounds? Helping you out is the least he could do."

"For one thing, the Japs sank his battleship, so I think he has a few other things to do at the moment," said Samuels. "More importantly, I know for a fact that at least two of his buddies on the *Nevada* were killed in the attack yesterday. I saw their bodies, and they weren't pretty. So you might cut him a little slack."

As they exited the hospital, Sherertz saw Mote sitting on a bench at the side of the hospital entrance. He was crying and looked wretched.

Sherertz walked over to him. "Johnny, what's the

matter?"

"I couldn't bring myself to tell Helen about what happened to Hal."

Sherertz put his hand on his shoulder and sat down next to him for a moment. "Johnny, don't worry about it. I'll take care of it."

Sherertz stayed with Mote a little longer, until he stopped sobbing, and then walked away leaving him crying quietly on the bench. Clutching his Bible with both hands, Sherertz looked just as miserable.

~~~

As the week went on, many of the officers and seamen were assigned search and recovery duty for bodies in the harbor. Today was Sherertz's turn to lead a group. He and the seamen picked out by the deck bosun's mate were just leaving the officers' dock, with Sherertz in the bow on the port side.

"Head for the south end of the 1010 dock and run slow next to the pilings," Sherertz said. "A lot of bodies have been surfacing there."

They started down near the wreckage of the USS *Shaw* and headed north next to the dock. All the men on the starboard side had long poles with gaffing hooks on the end. They moved slowly, barely greater than idle speed. The water was full of debris, covered with oil, and carried the stench of death.

"There's one up ahead," yelled one of the men in the

bow.

They pulled alongside, and Sherertz watched several of the men hook the body and begin to pull it out of the water. It looked gruesome—blue in color, eyes bulging, the whole body bloated to twice normal size. He could see several bullet holes in the back that oozed a black liquid as they moved the body, and one of the legs was missing below the knee. As the men slid the body up over the boat rail, Sherertz saw that the stomach was quite distended and stretched tight as a drum. "Be careful," he started to say, but it was too late.

"Whoooosh," went gas out of the stomach, followed by an explosion of foul smelling stomach contents all over the seamen pulling the body into the boat. Overcome by putrefied entrails, the seamen reflexively dropped the body, and its belly contents spilled out on their feet. Many of the men threw up. Some merely stared in horror. Several backed as far away as they could get and sat down with their whole body shaking. As they navigated the harbor, turning, speeding up and slowing down, the belly contents flowed back and forth on the bottom of the boat. Most of the men picked their feet up to avoid touching the foul liquid.

The men completed their circuit, ending at Aiea Landing, entirely in silence. Sherertz noticed their expressionless faces and the distant look in their eyes, as though wishing they were somewhere else.

When they arrived at Aiea Landing, he walked down to

the body on the deck and grabbed both of its hands. "I need a volunteer to help me move him," he said.

He looked around and saw no takers, and no one was making eye contact. Then he saw a large Negro man from the back of the boat stand up and walk toward him. Sherertz remembered that he was picked from the galley crew. Without saying a word, the man gently picked up one leg and what was left of the other, and he and Sherertz lifted the body up and carried it out of the boat.

"Thank you, Seaman," said Sherertz, holding out his hand and looking him right in the eyes. "I really mean it."

The seaman seemed taken aback and initially unsure of how to respond. After a brief hesitation, he reached for Jack's hand. "They call me Seaman Rivers, sir," he said as he shook Sherertz's hand. "The dead deserve our respect."

Ensign Sherertz never saw the man again, but he never forgot him.

~~~

Sherertz, Mote, and Drane were sitting in a bar the Friday night after the Pearl Harbor attack. None of them were talking. At least two dozen shot glasses were turned upside down in front of them.

Mote broke the silence. "I ran into Garfield today. He said that Jones, Uhrenholdt, and Booth were all killed when the *Arizona* blew up. He said he wouldn't be coming tonight." He started to say something else, but was unable to, as his face screwed up and tears came.

Sherertz replied. "I saw Joan Stern yesterday over at the Admiral Block Sports Arena. She told me that Charley wasn't even supposed to be on duty last Saturday night. The man who had the duty got transferred out earlier that day, so Charley volunteered. She'll be flying back soon to Albany, New York, so he can be properly buried. I couldn't think of anything to comfort her." The agony boiled up from deep inside Sherertz and tears came.

Sherertz, Mote, and Drane stayed together for a little while longer, and then, without another word, they stood up and left.

Drane and Sherertz did not meet again after they were reassigned new duties. Mote and Sherertz would see each other again, but they never talked about Pearl Harbor.

~~~

Sherertz arrived at Captain Scanlon's makeshift quarters on the Nevada. Sherertz didn't say anything and waited patiently to be recognized. Finally, Scanlon looked up wearily from the stacks of paper on the desk in front of him.

"I have filled out my reimbursement claim for you to look over and am ready for reassignment, sir," said Sherertz.

"At ease, Jack. Let's see what you've got." After quickly looking over the forms, he continued. "It looks all in order. I'll start it through channels." He handed Ensign Sherertz his orders. "You're being assigned to PT boats."

"You're kidding," said Sherertz. "No, you're not kidding. There's nothing else available, is there?"

Captain Scanlon smiled wearily and shook his head no. "I think you'll be pleasantly surprised. I've been out on one. It's been a pleasure working with you, Jack. Good hunting."

"Thank you, sir. Good luck on your next boat." Sherertz turned and left.

~~~

All total the one hundred ten minute, December 7th Japanese attack on Pearl Harbor killed 2,403 Americans (2,365 service men and women, 68 civilians), wounded 1,178 others, and sank or ran aground 18 ships, five of them battleships. On December 8th, the US declared war on Japan.

Figure 28. Mass grave burial ceremony on December 8th 1941 after Japanese attack on Oahu. These men were killed at Naval Air Station Kanoehe Bay. Largest number of men were buried at Punchbowl Cemetery. US Navy National Archives, history.navy.mil, Photo # 80-G-32854.

(see page 166)

Figure 29. Punchbowl Cemetery, December 7, 1942. AP Wire photo, NRW21055PL. Image can be found at www.wfirg.com/cemeteries/temporary and at www.AHeroAmongMillions.com.

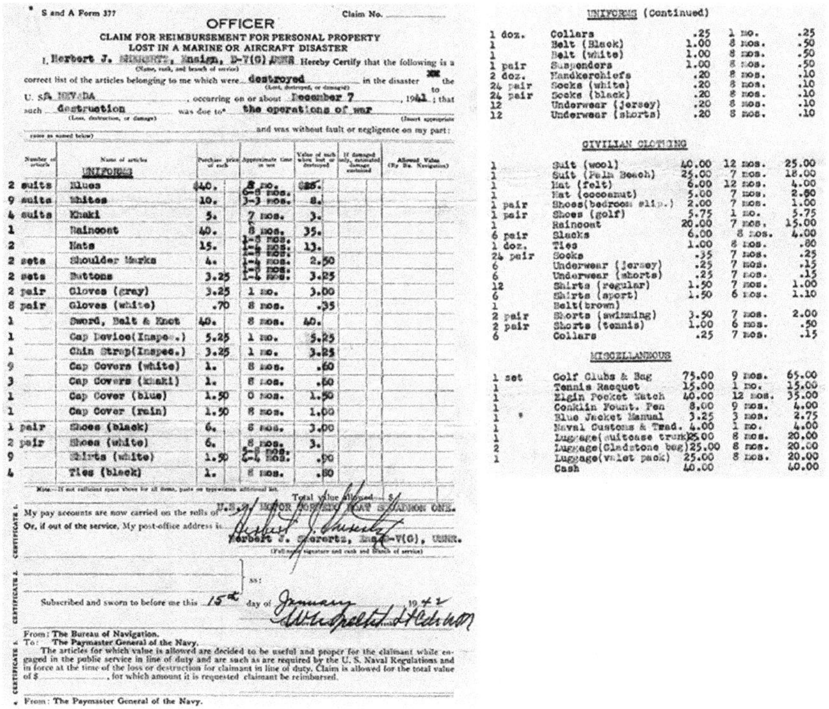

S and A Form 377 Claim No. ____

OFFICER

CLAIM FOR REIMBURSEMENT FOR PERSONAL PROPERTY
LOST IN A MARINE OR AIRCRAFT DISASTER

I, Herbert J. Sherertz, Ensign, D-V(G) USNR (Name, rank, and branch of service) Hereby Certify that the following is a correct list of the articles belonging to me which were destroyed (Lost, destroyed, or damaged) in the disaster to the U. S. NEVADA, occurring on or about December 7, 1941; that such destruction (Loss, destruction, or damage) was due to* the operations of war (Insert appropriate cause as named below) and was without fault or negligence on my part:

Number of articles	Name of articles	Purchase price of each	Approximate time in use	Value of each when lost or destroyed	If damaged only, estimated damage sustained	Allowed Value (By Bu. Navigation)
	UNIFORMS					
2 suits	Blues	40.	8 mo. 6-8 mos.	25.		
9 suits	Whites	10.	3-3 mos.	8.		
4 suits	Khaki	5.	7 mos.	3.		
1	Raincoat	40.	8 mos.	35.		
2	Hats	15.	1-8 mos. 1-4 mos.	13.		
2 sets	Shoulder Marks	4.	1-8 mos. 1-4 mos.	2.50		
2 sets	Buttons	3.25	1-8 mos. 1-4 mos.	3.25		
2 pair	Gloves (gray)	3.25	1 mo.	3.00		
8 pair	Gloves (white)	.70	8 mos.	.35		
1	Sword, Belt & Knot	40.	8 mos.	40.		
1	Cap Device(Inspec.)	5.25	1 mo.	5.25		
1	Chin Strap(Inspec.)	3.25	1 mo.	3.25		
9	Cap Covers (white)	1.	8 mos.	.60		
3	Cap Covers (khaki)	1.	8 mos.	.60		
1	Cap Cover (blue)	1.50	0 mos.	1.50		
1	Cap Cover (rain)	1.50	8 mos.	1.00		
1 pair	Shoes (black)	6.	8 mos.	3.00		
2 pair	Shoes (white)	6.	8 mos.	3.		
9	Shirts (white)	1.50	5-8 mos. 4-4 mos.	.90		
4	Ties (black)	1.	8 mos.	.80		

Note.—If not sufficient space above for all items, paste on typewritten additional list.

Total value allowed ... $

CERTIFICATE 1. My pay accounts are now carried on the rolls of U.S.N. MOTOR TORPEDO BOAT SQUADRON ONE.
Or, if out of the service, My post-office address is

Herbert J. Sherertz, Ens. D-V(G), USNR.
(Full name signature and rank and branch of service)

CERTIFICATE 2. ss:

Subscribed and sworn to before me this 15th day of January, 1942

CERTIFICATE 3. From: The Bureau of Navigation.
To: The Paymaster General of the Navy.
The articles for which value is allowed are decided to be useful and proper for the claimant while engaged in the public service in line of duty and are such as are required by the U. S. Naval Regulations and in force at the time of the loss or destruction for claimant in line of duty. Claim is allowed for the total value of $ ____, for which amount it is requested claimant be reimbursed.

CERTIFICATE 4. From: The Paymaster General of the Navy.

UNIFORMS (Continued)

1 doz.	Collars	.25	1 mo.	.25
1	Belt (Black)	1.00	8 mos.	.50
1	Belt (white)	1.00	8 mos.	.50
1 pair	Suspenders	1.00	8 mos.	.50
2 doz.	Handkerchiefs	.20	8 mos.	.10
24 pair	Socks (white)	.20	8 mos.	.10
24 pair	Socks (black)	.20	8 mos.	.10
12	Underwear (jersey)	.20	8 mos.	.10
12	Underwear (shorts)	.20	8 mos.	.10

CIVILIAN CLOTHING

1	Suit (wool)	40.00	12 mos.	25.00
1	Suit (Palm Beach)	25.00	7 mos.	18.00
1	Hat (felt)	6.00	12 mos.	4.00
1	Hat (coconut)	5.00	7 mos.	2.80
1 pair	Shoes(bedroom slip.)	2.00	7 mos.	1.00
1 pair	Shoes (golf)	5.75	1 mo.	5.75
1	Raincoat	20.00	7 mos.	15.00
6 pair	Slacks	6.00	8 mos.	4.00
1 doz.	Ties	1.00	8 mos.	.80
24 pair	Socks	.35	7 mos.	.25
6	Underwear (jersey)	.25	7 mos.	.15
6	Underwear (shorts)	.25	7 mos.	.15
12	Shirts (regular)	1.50	7 mos.	1.00
6	Shirts (sport)	1.50	6 mos.	1.10
1	Belt(brown)			
2 pair	Shorts (swimming)	3.50	7 mos.	2.00
2 pair	Shorts (tennis)	1.00	6 mos.	.50
6	Collars	.25	7 mos.	.15

MISCELLANEOUS

1 set	Golf Clubs & Bag	75.00	9 mos.	65.00
1	Tennis Racquet	15.00	1 mo.	15.00
1	Elgin Pocket Watch	40.00	12 mos.	35.00
1	Conklin Fount. Pen	8.00	9 mos.	4.00
1	Blue Jacket Manual	3.25	3 mos.	2.75
1	Naval Customs & Trad.	4.00	1 mo.	4.00
1	Luggage(suitcase trunk)	25.00	8 mos.	20.00
2	Luggage(Gladstone bag)	25.00	8 mos.	20.00
1	Luggage(valet pack)	25.00	8 mos.	20.00
	Cash	40.00		40.00

Figure 30. Ensign Sherertz's personal items destroyed by Japanese bombs in USS Nevada quarters. Lt. Cmdr. Jack Sherertz's World War II records (A,B).

UNITED STATES PACIFIC FLEET
BASE FORCE
P16-4/00/(7531)   U.S.S. ARGONNE, Flagship

SECOND ENDORSEMENT to   Pearl Harbor, T.H.,
PacFltPooling Off. orders   December 17, 1941
No.0-236

From: The Commander Base Force.
To :   Ensign H.J. Sherertz, D-V(G), U.S.N.R.

    1.   Reported at 1600 this date.

    2.   You will further proceed and report to the
Commander Motor Torpedo Boat Squadron ONE in accordance with
paragraph 1 of basic orders.

                                J.H. THEIS
                                XXXXXXXXXXX,
                            Staff Duty Officer.
- -
THIRD ENDORSEMENT   M.T.B. RON ONE   Pearl Harbor, T.H.,
PT/P16-4/00(jck)   December 17, 1941.

From: The Commander, Motor Torpedo Boat Squadron One.
To :   Ensign H.J. Sherertz, D-V(G), U.S.N.R.

    1.   Reported for temporary duty at 1730 this date.

Figure 31. Ensign Sherertz's reassignment orders to MTB Squadron One.
Lt. Cmdr. Sherertz's Military Record.

Figure 32. Military recruits flock to Roanoke, Virginia, Marine
recruiting station on December 8, 1941. Roanoke World-News clipping.
Can be found at www.AHeroAmongMillions.com

Sword and Pistol by his Side

The launch carrying Sherertz arrived at the portside gangway to the *Nevada*.

"I won't be long," Sherertz told the seaman piloting the launch. He walked slowly up the stairway. He knew this was likely to be his last trip to the *Nevada*, so he intended to savor it. He stopped several times along the way to look around. He could see that Pearl Harbor was bustling with activity and new purpose, felt at all levels of the Navy. "This will not happen again," he said to himself with resolve. Arriving at the top, he saluted the seaman guarding the deck. It was one of the men he and Drane had commanded.

"What's your business, Ensign Sherertz?"

"I need to go to my cabin."

"But, sir, you know it's underwater."

"I won't be long, Seaman."

Before the seaman could protest, Sherertz headed left toward the forward fourteen-inch gun turret. He wanted to see the damage, now that the smoke had cleared. A gaping hole below the gun barrels extended all the way down to where he could see water reflecting the light from above and spanning from one side of the deck to the other. He steeled himself to face whatever he was going to find below deck.

Then he walked aft past the seaman guard toward a familiar portal — the same portal he had gone down the day he arrived on the *Nevada*. He went inside, but this time he had no intention of sliding down the rails. He took his time. On the first level, the evidence of where the bombs had penetrated was ample. Looking around, he could see that there were already men on board beginning to clear out the wreckage. One of them was walking toward him.

"What's the plan for the ol' gal, sailor? Is she heading for the scrap heap?"

"Actually the rumor is that they're going to refloat her and send her to Seattle for a new face, sir."

"I hope you're right, sailor. She doesn't deserve to finish her career like this."

Sherertz headed toward the stairway. As soon as he turned to where he could have previously walked down, he saw the water. It came up to barely a foot below the ceiling of the second level. The strong smell of oil and burn triggered memories of the attack. He pushed them aside, took a deep breath, and stepped into the water. It was warm, as expected, but it was also quite oily, with a visible layer on top. When the water reached his neck, he ducked his head so he could get through the down portal. On the other side, his head was in an air pocket that was adequate for breathing, if he could tolerate the smell. He turned to his left, in the direction of Warrant Officers' Country, and began swimming slowly, looking carefully from side to side as he went. He found nothing much to draw his attention and soon reached the bulkhead that led to Warrant Officers' Country. It took a couple of minutes to find the portal. *Hope there's air on the other side or I'll be doing a quick turn around*, he thought to himself. Then he took a breath and swam down and through.

Coming up on the other side, he found absolute chaos. He had to swim around wires and pipes hanging from the ceiling and give wide berth to areas where the ceiling bulged downward, due to one of the bomb explosions. From a distance, he could occasionally see light from above that followed the bomb path down. Finally, he reached the second bulkhead leading to Junior Officers' Country. After a quick search, he found the portal and ducked under. Again he found more evidence of burning all over the ceiling. He found less debris blocking his path, so he headed to his left, keeping his hand on the bulkhead. When he reached the first officer's quarters, he swam along that wall, counting doors until he reached his quarters.

Without hesitation, he took a deep breath, ducked under the water, and headed into his room. When he came up, he had enough room to breathe, but just barely. Clothing was everywhere — on top of the water, under the water — some intact, most in pieces. He took a deep breath and dove down. It was like swimming in a washing machine. He ran into socks, underwear, a wool suit coat. He finally got tangled up in his raincoat and had to go back up for air.

The second time, he made it to the bottom and found his golf clubs, bent like spaghetti and burnt black, with a belt tangled up among them. On the third try, he reached his closet. He felt down the walls until he reached the bottom. There it was, his sword. He surfaced and put it above some pipes for safekeeping, then dove down again. His drawers were all shattered, but amid the debris, he found his pistol. By this time he was tired and winded. He held on to one of the pipes running across the ceiling while he caught his breath.

As he inhaled deeply, he realized that floating in front of him was a colorful paper umbrella, the kind you get in a mai tai or other fruity beach drinks. This one he had saved from the night Christopher had announced his engagement; he had made a toast with the drink that held the umbrella. Suddenly it all flooded back—his friends—Christopher, Davis, Jones, Booth, Uhrenholdt, Sterns—all dead, and the other men he knew or commanded who had died. *Why them, not me?* Deep down in the bowels of his dead ship, he had the urge to let himself sink under the water and join them. *Help me Lord, I really need you now.* Realizing his hands were starting to cramp because he was gripping the pipe so hard, he forced himself to slow his breathing and get control of himself. Unable to stand it any longer he let go, grabbed his sword and his pistol, and ducked out the doorway to his quarters.

Surfacing, he made his way back down the corridor,

past the bulkheads to the ladder, and climbed out. He noted his transformation: covered in oil, a pistol in his belt on one side, a curved sword on the other.

"Avast, me hearties, a pirate returns from the deep," said the seaman, grinning at the other seamen on watch duty.

Sherertz scowled, which completed his image as a pirate.

"I didn't mean anything, sir," said the seaman, but he couldn't keep from laughing.

Sherertz couldn't keep scowling any longer and cracked up. The laughter came from down deep and helped relieve some of the angst he felt a few minutes ago. Seemingly from nowhere, he found himself singing a song to himself. "Frog went a courtin', he did ride. Sword and pistol by his side, m-hm," sang Sherertz. It was a song his father had sung to him as a small child.

"Are you singing, sir?" asked the seaman with a smile.

"Thanks, Seaman, I needed that," said Sherertz with a big grin. He started toward the gangway.

"Sorry, sir, but you'll have to give me your pistol."

"Why is that, sailor?" asked Sherertz.

"The official line is that the salt water will have corroded the firing mechanism, but the rumor is that the brass are afraid we'll shoot up the Japanese on the mainland."

"They might be right," Jack said, knowing he had had some angry thoughts. So far, prayer had helped him get through, but some of his buddies had made out-loud threats.

"We'll get those Jap bastards, and we'll keep your gun until you come back. You might need it."

"I won't be back," said Sherertz, with a touch of sadness in his voice. He handed over the pistol and took one last, slow look at his boat. "Looks like the frog will go a courtin' without his pistol," he said to no one in particular.

"Attention, officer leaving the USS *Nevada*." All the men snapped to attention and held it until Sherertz dropped out of sight down the gangway.

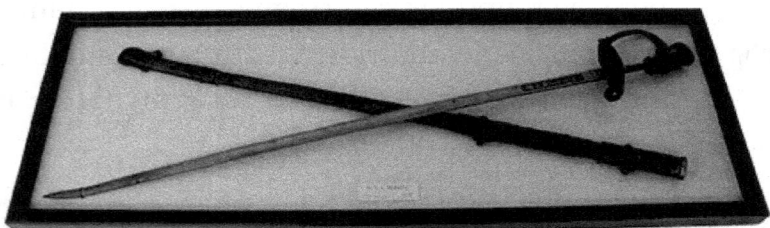

Figure 33. Ensign Sherertz's Navy sword. R. Sherertz collection.

Figure 34. S&W Model 10 Navy revolver. Wikipedia.

RON 1

Figure 35. MTB Squadron 1 PT Boat (PT-20). Navsource.org/
archives/12/05020.htm, US Navy photo (mod) 32475.

December 17, 1941

Sherertz arrived in front of a door that said "Lt. William Specht, CO RON 1". He collected himself for a moment and then knocked.

"Come in."

"Ensign Sherertz reporting for duty, sir," said Sherertz, as he opened the door.

Lieutenant Specht looked up to find Sherertz standing there expressionless. He looked exhausted and wore a wrinkled, stained khaki uniform.

"Welcome aboard, Ensign Sherertz. I see they haven't issued your new uniforms yet. Don't worry about it. We're a lot less formal on PT boats. I spoke with Captain Scanlon, and he said very good things about you."

At the mention of Captain Scanlon's name, Sherertz's face took on a strained look.

"PT boats will likely seem a real comedown after the mighty *Nevada*."

At the mention of the *Nevada*, Sherertz's face became downright rigid.

"I guarantee you that our PT boats can outrun anything in the Navy."

Sherertz stared straight ahead. He knew he was barely keeping it together.

"I know you lost a lot of friends at Pearl. How are you holding up?"

That was the last straw. Trembling and unable to speak, Sherertz almost collapsed. Lieutenant Specht got up and put his arms around Sherertz's shoulders. Sherertz began to sob. After a few minutes he stopped, and Specht stepped away.

"I would be shaken up, too, if I were in your shoes. You want a drink?"

Sherertz nodded.

Specht poured them both a shot of whiskey.

Sherertz threw down his shot with one gulp. By the time Specht finished sipping his drink, Sherertz had composed himself. "Please don't tell the men."

"Not a word. By the way, ensigns Mote and Morrison from the *Nevada* will be joining you."

"Well, that's something," said Sherertz, without emotion.

"I want you to head down to the dock, where you'll find your new boat, *PT-22*. Ask for Ensign Earl Fox, the executive officer of the boat. You'll meet your boat commanding officer, John Harlee, when he returns from breaking in another boat. I look forward to working with you."

Sherertz saluted, turned, and left without another word.

~~~

Ensign Sherertz walked down to the dock, heading toward several PT boats tied up to their slips. PT stood for Patrol Torpedo. At the beginning of World War II the US Navy held a "Plywood Derby" competition to see which US manufacturers would build the best PT boats. Higgins, ELCO, and Huggins won government contracts. PT boats were capable of speeds greater than 40 knots, powered by three Packard 1200-1500 horsepower diesel engines. Each boat had a hull made from two diagonal layers of mahogany planks with a glue-impregnated cloth layer in between, and ranged from 72-80 feet in length. Armament included two to four Mark 8 torpedoes, 2 M2 Browning .50 caliber machine guns, a forward twin .30 caliber machine gun, and later in the war, an aft 40 mm anti-aircraft gun.

Sherertz headed toward a 77 foot ELCO boat with *PT-22* painted prominently on the bow and just below the cockpit window. Before he could get close enough to walk up the boarding ramp, an ensign came running toward him. He started talking up a blue streak when he was barely in hearing range.

"Jack Sherertz, am I glad to see you," said Ensign Fox. "My name is Earl Fox. Welcome to the *Flying Deuces*. We put you in our squadron doctor's slot."

How ironic, thought Jack.

"We're calling you a signalman, but what we really need is someone who can instruct our landlubber gunner's mates. I hear you're just the man for the job." Fox closed the gap and began pumping Sherertz's hand. "Come on aboard, Jack, and I'll give you the five-cent tour. Listen up, guys. Ensign Sherertz knows everything there is to know about guns and is going to instruct all of you about how to use our .50 cals. We'll talk later, Jack. Gotta run." Fox left as quickly as he had appeared.

For a brief moment, everyone stared awkwardly at Sherertz. Then Sherertz walked over to the nearest pair of .50-caliber machine guns and looked carefully at the ammunition belt feeding from the right side. Next he checked out the firing mechanism. "Who are the gunner's mates?" he asked.

Several men raised their hands.

Sherertz turned to the gunner's mate nearest the gun. "Are you planning to fire this weapon?"

"Well . . . I think it's ready," said Gunner's Mate Blozis.

"If you're going to serve aboard the *Flying Deuces*, you need to give her namesakes the respect they're due," said

Sherertz. "On the *Nevada*, we called the .50 cals Deuces. They're excellent weapons, if used properly. With that in mind, there are a few things I might change before I fired this gun." Sherertz moved in and began to point them out. "The belt feed lever needs to be repositioned toward the right side of the cover. The pawl arm should point in the opposite direction: in this case, toward the left. Everything else is in proper position, although I think the gun will be more accurate and safer if you reduce the headspace. Too much headspace risks rupturing the barrel."

They all looked astonished. Then everyone started talking at once.

"Can you take a look at the firing pin?" asked Gunner's Mate Hill. "It seems to be wearing unevenly."

"When it gets hot, sometimes it jams in the third position," said Blozis.

"Sometimes it gets stuck in position one," said Hill. "I think the firing pin spring is the problem."

"All right, let's strip it down and see what we can figure out," said Sherertz.

~~~

The *Flying Deuces* headed at full speed toward a target barge with Gunner's Mate Blozis manning the forward .50-caliber machine gun. He hit the target on the first straight-on pass. Then Fox, the executive officer — usually referred to as XO — began evasive maneuvers and brought the boat back in a tight turn. Blozis missed the target entirely. The boat throttled down to a slow idle.

"What happened?" asked Sherertz.

"I lost my balance in the turn," said Blozis.

"Try leaning into the turn and leading the target," said Sherertz.

As Sherertz helped reposition Blozis, it was clear that Blozis looked sick. Sherertz pointed him toward the side. "Gunner's Mate Hill, do you want to take a try?" Sherertz asked.

But Hill was already puking over the side of the boat.

"I think target practice is over," said XO Fox. "Let's head back to dock. You two strip down the guns and clean and oil them."

~~~

Blozis and Hill were talking to Fox on board the Flying Deuces at the dock later that afternoon.

"First-rate cleaning, gentlemen," said Fox.

"We didn't do it, sir," said Blozis. "Ensign Sherertz did."

"We couldn't get him to talk," said Hill. "We invited him to go out with the crew and have a few beers tonight, but he wasn't interested."

"Give him time," said Fox. "He lost a lot of friends at Pearl."

"Sir, doesn't he know that we named the *Flying Deuces* after the Laurel and Hardy movie, and the obvious double twos?" said Hill.

"I think I like his interpretation of the *Flying Deuces* better," said Fox.

Figure 36. Life on a PT Boat. Roanoke World-Newsclipping (February 4, 1942). Can be found at www.AHeroAmongMillions.com.

Figure 37. Lt. Bulkeley PT Boat Raid. Roanoke World-Newsclipping (January 21, 1942). Can be found at www.AHeroAmongMillions.com.

Oahu to Midway

May 25, 1942

"Admiral Nimitz has intelligence that the Japs are likely to attack Midway between late May and early June with everything they've got," said Lieutenant McKellar, new CO of RON 1, addressing all the RON 1 officers assembled on the dock. Midway is an island that is a US territory and part of the same volcanic island chain as the Hawaiian Islands. "If the Japanese capture Midway, they'll be coming after Pearl Harbor next. We have a surprise in store for them, but we need to get there in a hurry. We'll be providing sea support for the pilots and trying to sink any wounded Jap ships. We're leaving immediately and are ordered to travel as fast as we can without damaging the boats. Good hunting."

The officers immediately hurried to their respective boats where their crews already had the boats idling and ready to go. All of the squadron immediately began to cast off. One by one, the boats left their birth and headed out to sea.

Ensign Sherertz and Lt. George Matteson were talking on the bridge of *PT-29*. Ensign Sherertz had been moved from third officer of *PT-22* to XO of *PT-29*.

"All right, XO Sherertz, are we ready to go?" asked CO

Matteson.

"We are, but we're going to have to travel at near full throttle most of the time to get to Midway in four days," said Sherertz. "It's about thirteen hundred miles. If the weather holds, we can do thirty knots for five hours straight and then refuel. That means we'll have to travel ten hours a day to get there in time."

"Agreed," said Matteson. "Listen up, everyone. We're hauling ass to Midway. Stay leeward of the islands. If the waves are low, open her up. Don't push it if you encounter big waves. We want to get all our boats there in one piece. We're not going to have time to fix any boats once we get there. No one's ever gone this far, this fast, in a PT, so everyone look sharp and keep your eyes open for Jap planes."

"What's our ETA, Skipper?" asked one of the gunner's mates.

"May twenty-eighth."

"Yes, sir. Balls to the wall," said one of the motor mechs.

The motor mechs were slapping each other on the back and fairly jumping up and down with excitement as they headed below. It wasn't often that they were ordered to go at maximum speed. Everyone else moved to his assigned positions, happy to be doing something besides drilling twenty-four hours a day.

"Sherertz, you and I will rotate driving the boat," said Matteson. "Make sure all the men are getting some sleep so everyone stays fresh."

~~~

It was now late afternoon. Sherertz climbed down the ladder from the bridge to the chart room and then walked forward to the crew's quarters. On four stretcher-like beds hanging from the ceiling, four men slept facedown with both hands holding on to the side edges and one foot hanging over each side. As he watched, the boat flew over a wave and slammed back down. Each crewmember held on in his sleep and didn't miss a wink. With all well there, Sherertz wandered over to the engine room. Everything seemed in order, so he headed back up to the bridge.

~~~

Figure 38. USS *Ballard* (1943), seaplane tender, Wikipedia.

The first night, RON 1 stopped at Necker Island after dark and was refueled by the USS *Ballard*, a seaplane tender commandeered for this mission. To refuel more boats at once, some of the refueling was done manually using hand cranks from barrels carried on the *Ballard's* deck. All told, it took several hours. After refueling, they set the watch, and everyone else crashed for some rack time.

~~~

The next day they got an early start and made good progress through calm seas to reach the *French Frigate Shoal* by late afternoon. While the *Ballard* and some tuna boats initiated refueling, all officers came on board the *Ballard* to talk about their upcoming mission at Midway. The meeting was short, and soon they were back out on deck for a break before dinner with the men. Sherertz was standing near Matteson looking out at the reefs when a stranger in khaki pants and a baseball cap approached.

"Sherertz, I am Commander John Ford. I direct movies, like *Stagecoach*, and I'm making a documentary about the attack on Pearl Harbor. Some of the other officers told me that you were at Pearl Harbor on the USS *Nevada*, and I was wondering if I could ask you a few questions."

"No, thank you," replied Sherertz, without hesitation.

"I spoke with McKellar, and he cleared my talking with you," said Commander Ford, as he studied Sherertz's face. Sherertz was clearly not happy. "It will only take a few minutes. If it helps, my orders come from Admiral Nimitz."

"I don't like talking about it, Commander Ford. Why don't you talk with ensigns Mote or Morrison? They were both on the *Nevada*."

"I already did."

Sherertz looked at Matteson, who shrugged his shoulders.

"Okay, I'll answer a few questions, but I won't talk about the men. By the way, why are you here? Surely not to interview me."

"I'm here to film *The Battle of Midway*."

Boom!

An explosion behind them lit up the night. One of the tuna boats had hit a mine. The other tuna boats tried to put space between them and the burning wreck that remained.

Boom! Another mine explosion rocked one of the remaining tuna boats, and everyone watched helplessly as it burned. Both boats sank in a few minutes. Instead of sleeping, as they all desperately needed, they sadly pushed through a funeral service in the dark.

~~~

The next day, the good weather continued, and they made it to Lisianski Island by late afternoon and refueled again. The following day they made it to Midway without difficulty.

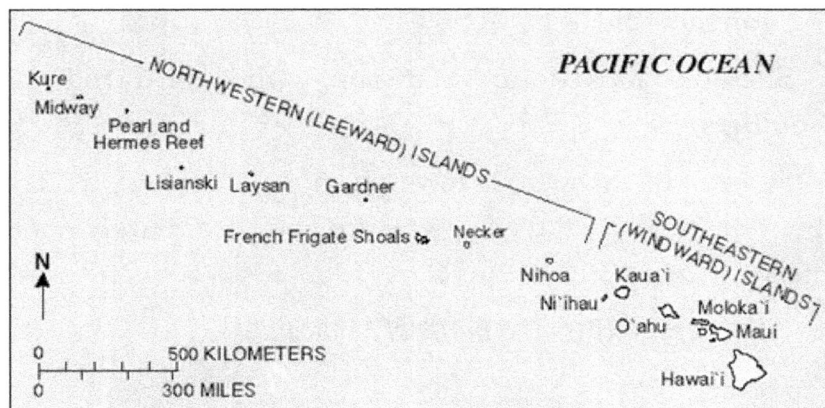

Figure 39. Hawaiian Islands. Wikipedia

Midway

Figure 40. Midway Islands, November 1941. Sand Island above, Easter Island below, Brooks Channel in between. US Navy National Archives, history.navy.mil, Photo # 80-G-451086.

May 28, 1942, Sand Island

Lieutenant McKellar's PT boat arrived at the Sand Island Midway dock right at sunset. He stepped off the boat to meet Marine Major Benjamin Norris. "Lieutenant McKellar, CO of PT RON One, reporting for duty." He saluted briskly.

"At ease," said Major Norris. "Welcome to Midway,

Lieutenant McKellar. Call me Ben. We're pretty informal here. I'll have Second Lieutenant Ringblom show you to your quarters. Listen, McKellar, any chance you have a chaplain? We usually have a church service on Sunday at zero eight hundred hours, but we had to ship our chaplain back to Pearl due to an attack of colitis. With the battle coming up, I think it's especially important to have the church service as usual."

"We don't have a chaplain, but Ensign Sherertz used to teach Sunday school before the war."

"He'll do nicely."

"Oh, and by the way, there's a poker game every Friday night in the Marine officers' quarters. In case you've lost track of time, that's tomorrow night."

~~~

During the day Friday, the PT boaters explored the Midway Atoll and surrounding waters, getting to know the shoals and sandbars. Later, while the mechs were overhauling all the boats, Sherertz and Mote decided they needed some R&R. They walked into the Marine officers' quarters after chow Friday night, and found Second Lieutenant Allan Ringblom with some of his Marine buddies.

"Yo, Marine grunts, where's the poker game?" Sherertz asked.

Ringblom came right over to them. "Greetings, Navy scum," he said with a big smile. "My name's Al. I hope you come laden with gold, as we like big pots around here. Whom do we have the pleasure of fleecing?"

"My name's Jack Sherertz, and this is Johnny Mote."

They all shook hands.

"Come meet the guys and cash your dough for chips," said Al.

Hours later, as the game was breaking up, Al walked over to Sherertz. With a sheepish grin, he looked at Sherertz as though not sure if he should ask, and then asked anyway. "Listen, me and some of the other pilots need to get over to Eastern Island for target practice every day. A garbage scow has been ferrying us back and forth. It really stinks. Now that your group is here, do you think someone could give us a ride over?"

"Sure, Al. Having friends in high places could come in handy."

"Are you kidding? The guys won't believe it! We will be eternally grateful, even more grateful than accepting your monetary donation to our pockets tonight. Can someone take us over at thirteen hundred hours tomorrow?"

"No problem. Look for PT- 29 at the dock. That's my boat."

"Everything is jake today!" said Ringblom.

~~~

Sunday, May 31, Sand Island

Sherertz stood in the mess hall in front of a group of

men, reading a psalm from his Bible. With their heads bowed, the men said one final "Amen," raised their heads, and began to exit the room.

Second Lieutenant Ringblom came up to Sherertz as he walked out the door. "Jack, thanks for giving us a church service on such short notice," he said. "Do you think you can hang out with us for an hour after we get to Eastern Island today?"

"Sounds very mysterious, Al."

"It'll be fun, I promise you."

~~~

May 31, 1942, Eastern Island

Ringblom and Sherertz got off PT- 29 onto the dock at Eastern Island. After a short walk, they arrived in front of a two-cockpit plane.

"Jack, this is my baby. I call her Alice. She's an SB2U-3 Vindicator. A little rough in high winds, but most times I can drop a bomb on any target there is to hit. Wanna go up? I need to do my daily dive-bombing practice, and Major Norris sent my gunner on an errand."

The SB2U-3 Vindicator was a single propeller monoplane developed after a 1934 US Navy call for monoplanes and biplanes that could take off and land from an aircraft carrier. They were manufactured by Vought and powered by a 825 hp Pratt & Whitney engine able to reach a cruising speed of 146 miles per hour. They were designed as divebombers to carry a 500 or 1000 lb bomb.

Figure 41. SB2U-3 Vindicator taking off from Eastern Island
ust before the Battle of Midway. Wikipedia.

"Al, I'd love to. By the way, do we have to worry about running into a bunch of gooney birds?"

"Nah. As long as we don't spook them, they're happy on the ground. Here, put on this helmet and a flight jacket and a pair of goggles. Oh, and one more thing. Hang on to this."

"What is it?"

"It's a practice bomb. After you see me drop one, I'll let you see if you can hit our target. Climb up that ladder and step on the wing. I'll show you how to get in the cockpit once you're up there."

The Navy AN practice bombs used by SB2U-3 pilots were typically eight inches long, weighed 3-4.5 pounds, and were made from either zinc-alloy (MK-5), cast iron

(MK-23), or lead-antimony (MK-43). The MK-5 was designed to practice on ½ inch armored boat deck targets and was what Ringblom had given to Sherertz for their target practice today. If desired, it could be filled with a 10-guage shotgun shell or fluorescein dye to show where the target was hit.

~~~

Vindicator Alice taxied down the runway and took off. It spiraled steadily upward, gaining altitude, then leveled off. Below, Sherertz could see the Midway Atoll — eight miles wide, Sand Island with the PTs docked, and Eastern Island with one plane taking off, followed immediately by another one landing. Ringblom looked totally at home in the pilot's seat. Sherertz sat in the gunner's seat behind him facing the tail of the plane. Initially nervous he gripped the plane fuselage way too tight, but after a few minutes of staring out the cockpit window, he began to enjoy himself.

"All right, Jack," said Ringblom. "Now the fun begins. "Slide your cockpit open and look out the side to your left."

Sherertz opened his canopy as instructed and then turned in his seat, ninety degrees to his left. Fortunately the tail machine gun was being serviced so it didn't obstruct his movement.

Without further notice, Second Lieutenant Ringblom went into a steep dive. "That's our target below on Spit Island, that big circle."

"I see it."

"I'm going to slide my cockpit canopy open and drop the practice bomb on the target. Watch and learn."

Seemingly at the last possible moment, he let the bomb go. Sherertz watched out the window as Ringblom pulled up sharply. There was a burst of sand into the air, almost dead center of the target.

"Nice shooting. But can you do that with a Zero on your tail?"

"I sure hope so. Now it's your turn. Hold the bomb out the window once we go into our attack dive. Don't throw it. Just let it go. Here we go."

Sherertz held the practice bomb out as instructed.

"Okay, Jack. Bombs away."

Sherertz let go of his bomb and looked back. His bomb missed the target by five yards.

"Nice going. My first try missed by ten yards. I'm going to set her down and let you out. I have to do touchdowns, and I don't think you'll find that nearly as exciting."

~~~

June 1, 1942, 0630 hours

PT- 29 pulled slowly away from the dock on the eastern side of Sand Island with her engines fully muffled. The boat headed toward Brooks Channel, which ran down the middle between Sand Island and Eastern Island, bracketed by Gooney Island on the Sand Island side of the channel and Spit Island on the Eastern Island side of the channel. As they turned into the channel, the sun was just beginning to come up in the east over Spit Island.

"Looks like another spectacular sunrise," said XO Sherertz. "I'm going to enjoy this last bit of peace and quiet before the day gets rolling."

"It would be even nicer if you'd stop flapping your gums," said CO George Matteson with a smile.

Before Sherertz could respond, they heard the distinctive sound of .50-caliber machine-gun rounds from Gooney Island, "Tat, tat, tat, tat, ...", followed immediately by, "Crack, crack, crack, crack, ...," as rounds tore through the wooden walls of the PT boat.

"Cease fire or there'll be hell to pay!" yelled Matteson.

"Sorry, thought you were our scheduled target practice," yelled a very cautious voice from the shore. "We couldn't make you out with the sun behind you. Everyone okay?"

"If anyone's hurt, I'll hunt you down and court-martial you personally!" said Matteson.

Marines from the machine-gun embankment were seen running for cover. With each step they took, Gooney bird guano floated into the air in clouds glowing bright orange in the rising sunlight.

"Anyone hurt?" asked Matteson.

"All's well, Skipper," said Sherertz.

"Damage report," said Matteson.

Voices from around the boat began reporting. "None in the engine room. None in the forward quarters. All guns intact. Wheel and instruments fully functional. Smoke generator and torpedo housings untouched."

"How dare they shoot up my boat!" yelled Matteson to no one in particular. He stomped from the bow, climbed up to the day hatch, and disappeared below.

Everyone got out of his way, and all remaining crewmen on deck stared defiantly toward Gooney Island.

~~~

June 1, 1600 hours

Ringblom approached PT officers' quarters and knocked on the door. "Is Jack Sherertz here?"

"Who wants to know?"

"I have an olive branch from our Marine squad on Gooney Island."

Ensign Sherertz came out the door. "Keep your voice down, Al. Our skipper is still really pissed."

"I would be too, if friendly fire shot up my plane. Anyway, the guilty parties have sent you their belt knives as a peace offering and apology. When your skipper settles down, let him know that our CO ordered them to clean out all the latrines on the island. Anything else I can do for you?"

"Take a picture of their smiling faces covered with shit. I think that might cheer up our skipper."

Figure 42. Marine knife, sheath, box, and leather cord presented as an apology to Lt. (jg) Sherertz on Midway Island. Jack Sherertz collection.

Japanese Attack Midway

Figure 43. PTs and Zeros, Battle of Midway, Griffith Baily Coale oil painting. US Navy National Archives, history.navy.mil, #32, 88-188-AF, (original in color).

June 3, 1942, 0930 hours

PT Squadron One officers assembled in front of the barracks facing Lieutenant McKellar.

"We just received word from CINCPAC that the Japanese fleet has been sighted seven hundred miles northwest of Midway," said Lieutenant McKellar. "At their present course and speed, it looks like they'll be here tomorrow. Everyone recheck all your weapons, engines — everything."

The officers cheered.

"Ensign Sherertz, I wanted to tell you that your CO,

George Matteson, volunteered to lead *PT-29* and *PT-30* tomorrow to check out whether the Japs are moving on Eure Island. We're pretty sure they'll hit Midway, but we're not sure about Eure. While I know I just made you exec on *PT-29*, I talked it over with the other COs, and we all want you to help out Earl Fox and John Harlee on *PT-22*. Mote and Morrison will be on boats here at Midway, and I thought you would want to be here, too. This is your chance to get even for Pearl Harbor."

"Give 'em hell, Jack," said a voice from the back.

"An eye for an eye!" yelled another officer. The room was getting pretty rowdy.

Sherertz thought for a moment and then said, "Thank you, Lieutenant McKellar, but I'll have to decline. My beliefs do not allow for revenge. My boat has been ordered to Eure Island, and I will go with it."

It suddenly got very quiet. Then the assembled officers began talking to each other in low voices.

McKellar looked hard at Sherertz for a moment, as if trying to decide how to respond. "I respect that, Jack. All of you get what sleep you can, as you may not get much for a while after that."

"Let's see how well the Japs sleep after we shoot a torpedo up their ass," said one of the assembled officers.

His comment was met with laughter and more cheering.

"I'll see you in the morning right here at zero four hundred hours sharp," said McKellar. "Pleasant dreams."

~~~

June 4, 1942, 0300 hours – reveille

All the barracks on Sand Island emptied in a matter of minutes, as most soldiers were already fully dressed and wide awake. Marines ran to their gun positions, and Marine pilots jumped on board *PT-29*, which took off in a hurry for Eastern Island. After the pilots exited the boat, it headed toward Eure Island. Matteson and Sherertz had been told to monitor the radio continuously for anything related to the anticipated attack on Midway and to keep the CO appraised of any action on Eure.

The rest of the PTs dispersed out into the Midway Atoll lagoon or just offshore and waited. Finally at 0400, things began happening.

"There go Red Park's Wildcats . . . and now the Buffaloes," said XO Fox. "Oh baby, there go the flying fortresses to blow those Japs out of the water. This time it's our surprise."

~~~

Midway radio crackled. "This is Lieutenant Chase calling Midway Base, over."

"We read you loud and clear," said Major Parks.

"Many enemy planes heading toward Midway," said Lieutenant Chase.

Air raid sirens cranked to their full wail. Everyone manned their guns and prepared for action. The last of the Buffaloes and Wildcats took off, followed by a group of

TBFs and B-26s.

"Tallyho!" said Captain Carey over the radio. "Hawks at angels twelve."

~~~

"Lost contact with Major Parks and the rest of my squadron," yelled Lieutenant Daniel Irwin. "Zeros on my tail. Coming in red-hot. Need ground support."

"That's us, men," said CO Harlee. "Prepare to fire. There they are. Wait until they're in range. Now, give 'em hell!"

The *Flying Deuces* erupted with all guns aimed skyward at the incoming Japanese planes. From all over the lagoon, the other PTs opened fire, as did all the antiaircraft shore batteries. The sound was deafening.

Within seconds, the first few Japanese planes disintegrated, torn apart by the furious barrage of shells. One crashed into the lagoon, narrowly missing two of the PT boats. But soon some of the Japanese planes got through. Bombs began to fall, and some of the planes strafed the islands. The US forces had left a plane without an engine on the runway as a trap. When a plane came in to strafe it, they shot it down. One of the Jap planes dove and flew upside down just above the Eastern Island runways and was promptly splashed for it's showboating. Some of the bombs took out the powerhouse near the runway. Sparks and flashes of light flew in all directions, but only minimal damage was done to the runways. On Sand Island, three oil tanks and a seaplane hangar were set afire, and most of the buildings were destroyed. A Marine detachment in a dugout was bombed. No movement was seen afterward.

Commander Ford's crew had set up on top of the powerhouse on Sand Island. They filmed furiously during the whole battle—most of the filming done by Ford himself. The sky turned black with the smoke from the fires on the land and in the air.

"There's Irwin's plane," said CO Harlee. "Don't shoot it. Fox, move us closer to Eastern Island so we can keep the Japs off his tail."

"Gunners, concentrate your fire on the Zero closest to Irwin's plane," said XO Fox. "All right, it's on fire."

In spite of the planes chasing him, Lieutenant Irwin managed to land and pull into a hangar to rearm. The Zero right behind him tried to gain altitude but crashed in the trees on Sand Island.

"Out of gas . . . going to ditch . . . two miles northwest of Midway," said Second Lieutenant Ringblom.

"*PT-26*, you go get Ringblom," said CO McKellar, "and *PT-20*, go with him and look for survivors. The rest of you keep firing at the planes."

Minutes later, the Japanese planes turned and left.

At 0715 the all-clear siren sounded.

"Sparks, get Matteson on the box," said McKellar. Minutes later, the radioman handed McKellar the microphone. "Matteson, what's going on at Eure?"

"Dead quiet here, sir," said Matteson.

"Well, double-time it back to Sand Island," said McKellar. "You're needed to put out fires and look for survivors."

~~~

Jack and Fox were directing some men fighting a fire when McKellar stepped out of Midway headquarters and hurried over to them.

"Sherertz and Fox, *PT-20* and *PT-26* are coming back with survivors. Go see if they need any help."

"C'mon, Foxy, let's go find out who they picked up," said Sherertz.

They ran to the docks in time to see two Marine aviators jumping off *PT- 26*. Ringblom turned toward them with a big grin on his face.

"We didn't get many of them, but we kept them busy," said Ringblom. "Private Gene Webb, my gunner, I'd like you to meet two of my PT buddies, XO Jack Sherertz and XO Earl Fox."

Sherertz hugged them both and slapped them on the back. "I'm sure glad to see you back in one piece," he said.

"Dump your gear," said Fox. "There's a lot to clean up."

"Any word on casualties?" Ringblom asked.

"At least ten in the air and ten on the ground," said Sherertz. "The rest of our PTs are out cruising the lagoon and the bay looking for others. We're going back out to help them. See you later."

~~~

June 4, 1900 hours

At the parade ground in front of Midway headquarters, Lieutenant McKellar addressed the assembled RON 1 officers.

"We've just received word from CINCPAC that our carrier planes have disabled four Japanese carriers one hundred seventy miles northwest of us. We've been given orders to sink any that we find afloat."

All the officers cheered.

"The other part of the mission is to find any pilots that are downed," said McKellar. "Keep your eyes open. Dismissed."

The men ran to the dock and boarded their PTs.

~~~

The Japanese had sent thirty-six Kate torpedo planes, thirty-six Val dive-bombers, and thirty-six Zero fighters to attack the Midway atoll. All but eleven returned to their fleet. Midway sent up sixty-six aircraft. Nineteen were shot down and thirteen pilots were killed. Only eleven men of the Midway ground forces were killed. No casualties occurred on the PTs.

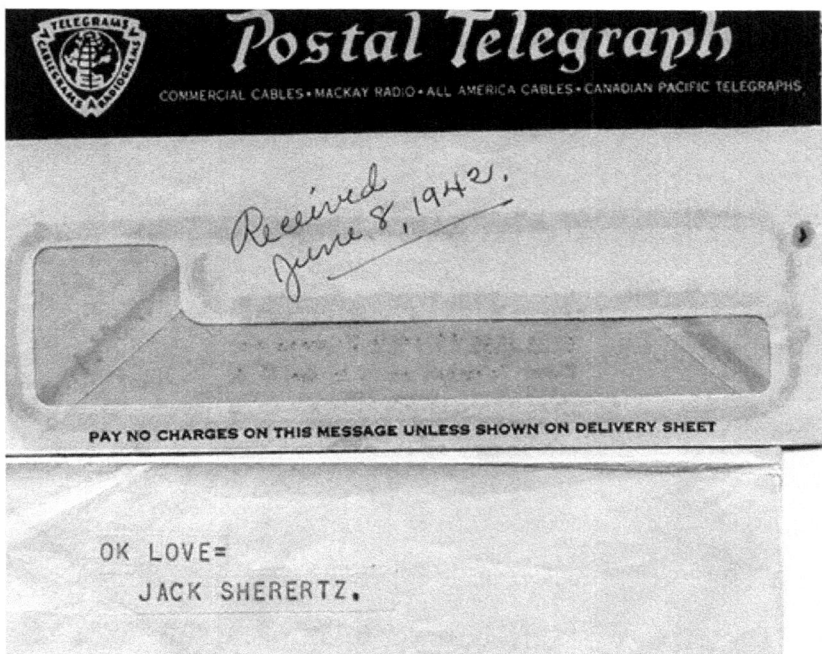

Figure 44. Jack Sherertz's Telegram to let his parents know he's alive.

Burials at Sea

June 5, 1942, 0700 hours

PT- 29 had just tied up at the dock at Sand Island; Sherertz and Matteson came off last.

Ringblom approached. "Any luck, Jack?"

"Oil and flotsam, but no ships and no pilots. I think the Jap ships sank before we got there. Anything happen while we were gone?"

"An enemy submarine shelled us for a few minutes, but the shells all landed in the lagoon. We chased it off with return fire and haven't heard from it again. Listen, Jack, can I talk to you a minute in private?"

The two of them walked along the dock.

"What's on your mind, Al?"

"Jack, we tried to bury one of our eleven dead pilots and kept running into water before we could dig deep enough. It's just not right to bury them like that. We all liked the church service you did for us and thought you might be able to come up with something better. Any ideas?"

"Marines are part of the Navy. Why don't we give them full Navy honors and bury them at sea? It's the way I'd like to be buried if I were in their place. If you think that will work, I'll take care of the details."

"Jack, that will be perfect."

The two shook hands.

"I have to go," Ringblom said, choking up. He turned and walked away.

~~~

McKellar walked up to Sherertz and held out his hand. In it were lieutenant junior grade stripes and a pin. "Jack," he said, "we usually do this more formally, but things have been a little hectic. This came in just before the Japanese attacked. With your change in rank and recent performance, I'm promoting you to boat captain. You'll take over *PT- 29* as CO. Good hunting." As quickly as he had spoken, he was gone.

~~~

June 6, 1942

Each of the eleven PTs was docked at Sand Island. All the men on board, dressed in their cleanest uniform, stood at ease.

Commander Ford approached *PT- 29* and addressed Lieutenant Junior Grade Sherertz. "I spoke with your CO about coming on board *PT- 29* and filming the funeral. He said that you were in charge of the ceremony and that it was up to you."

Sherertz came down to the dock and stared at Commander Ford a minute before speaking. "I know your orders come from Admiral Nimitz, but listen carefully," he said. "I'll let you come with us under the following conditions: your group is confined to the stern, and you cannot say a word from start to finish. If I sense at any time that you have taken advantage of our generosity, I will personally push you and your cameras off our boat, and you can film the sharks instead."

Without any further discussion, Sherertz returned to his boat.

On shore, Army and Marine personnel formed a large circle surrounding eleven coffins draped with American flags.

The base Commander, Col. Harold Shannon, addressed the assembled group. "Honor guards, come forward and hoist your Marine aviator. Forward, march. Halt. Transfer each aviator to his boat. Marine escorts, board your burial ship. To our departed brothers, may God be with you. You have not died in vain. Now, a moment of silence . . . Dismissed."

Each PT boat pulled out slowly and then fell in line until they were all in single file heading out to sea. The last boat was *PT-29.*

After several minutes, Ringblom spoke quietly to Sherertz. "How far out are we going?"

"About fifteen miles, so we're clear of sharks."

No further words were spoken. At the agreed-upon location, the first PT in line began to turn gently to port, and the other boats followed until they formed a large arc. Engines were cut to idle and muffled. Honor guards on the bow of each PT consisted of one line of three PT men and one line of three Marines. They were arranged adjacent to the port side of the bow facing each other. A flag-covered coffin rested between them on a large piece of plywood.

On his boat, Sherertz was in the middle of the bow, facing the honor guard. On the starboard side of the bow was a three-man firing squad. CO Matteson and a bugler were on the bridge.

Sherertz nodded to Matteson. Matteson raised his hand, and the motion was duplicated on each boat down the line. The ceremony began.

"All hands bury the dead," said Matteson. "Attention . . . Parade rest."

"Oh God, we pray thee that the memory of our comrade, fallen in battle, be ever sacred in our hearts," said Sherertz, "and that the sacrifice which he offered for our country's cause may be acceptable in thy sight. And may this Marine find repose with the sea in which he gave his life that we might live. Into thy hands, oh Lord, we commend the soul of thy servant departed, now called unto eternal rest, and we commit his body to the deep. Now all please join hands. Lieutenant Junior Grade Ringblom will say a few words."

"Major Floyd Parks was a fellow Marine and a good friend to all of us. Our hearts are with you, and we pray for your salvation. Amen."

"Attention!" said Matteson.

"To the deep we commend your spirit," said Sherertz.

Matteson raised his hand and looked to the other boats. When the hands of all the COs were raised, he lowered his, and the burial at sea concluded. The honor guard on each PT raised the plywood under the casket, and the casket slid out from under the flag, overboard. The CO pointed to the firing squad, which fired three volleys. After taps were played and the flag was folded by the honor guard and handed to the CO, each PT got underway and returned to Midway.

~~~

Upon return, Commander Ford came over to Sherertz. "Sherertz, there's someone I want you to meet," he said.

Standing on the dock was a casual-looking man in an officer's uniform with lieutenant Commander stripes.

"Sherertz, this is Lieutenant Commander Griffith Baily Coale," said Ford. "He's an artist assigned to the Navy to record the war through his paintings. He's heard that there's going to be a similar funeral for Japanese pilots, and he would like to go along."

"Pleased to meet you, Lieutenant Commander Coale," said Sherertz.

"The pleasure is all mine, I assure you," said Coale. "Do you mind if I tag along? I promise I'll stay out of the way."

Sherertz looked at Ford and back at Coale. "Sure, why not? If Commander Ford can stay out of the way, I'm sure anyone can." Sherertz sent a big grin in Ford's direction, which was immediately returned.

Figure 45. US Marine pilot burials at sea after Battle of Midway.
Jack Sherertz collection.

Figure 46. Sinking Sun (Battle of Midway, PT boats carrying Japanese pilots out
to sea for burial), Griffith Baily Coale oil painting. US Navy National Archives,
history.navy.mil, #28, 88-188-AB, (original in color).

Trigger Happy Destroyer

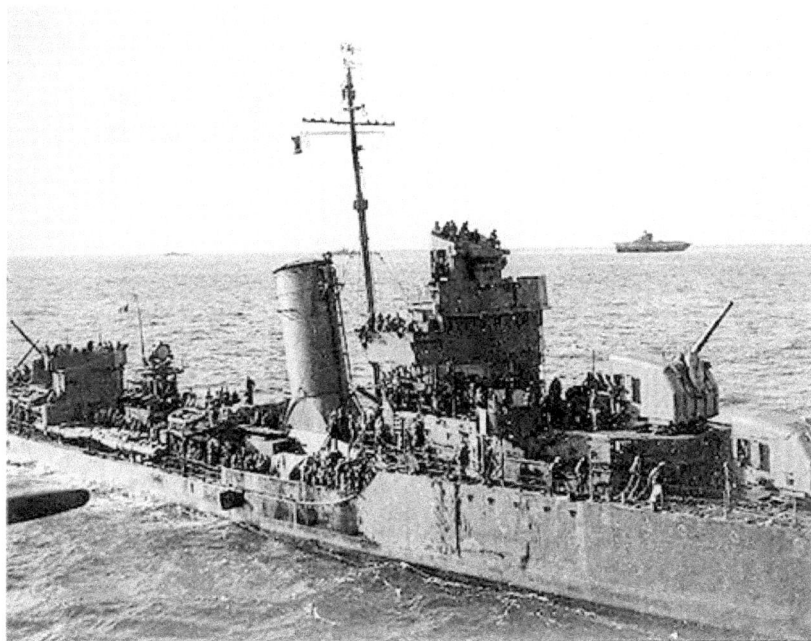

Figure 47. DD USS *Benham* with Yorktown survivors. Yorktown in distance on right. US Navy National Archives, history.navy.mil, Photo # NH 95574.

June 7, 1942

The naval victory at Midway had been decisive. The US lost only two ships, the carrier USS *Yorktown* and the destroyer USS *Hammann* and 340 lives, compared to the Japanese loss of four carriers (*Akagi, Kaga, Hiryu*, and *Soryu*), a heavy cruiser (*Mikuma*) and 3,057 lives. On June 7, US Task Force 17 broke up, with three destroyers joining

213

Task Force 16 to remain in the Midway area, and the remaining ships heading back to Pearl Harbor. Task Force 16 destroyers (including the USS *Balch*, USS *Phelps*, USS *Ellet*, USS *Benham*, USS *Maury*) and Midway-based PT boats were ordered to patrol the waters around Midway, looking for downed US pilots and the Japanese navy. No information was available as to whether the Jap ships had left the area, so tensions were high.

~~~

Early AM on June 8th, Lt. j.g. Sherertz ordered *PT- 29* to head northwest of Midway along the course taken by US planes and ships in the Battle of Midway, only a few days previously. The seas were rough, with swells up from fifteen to twenty feet high, but visibility was good. The men were looking for any damaged Japanese vessels, with orders to engage, or for any downed US or Jap pilots or sailors. The whole crew was on alert, with all men on deck and all guns manned.

Sherertz stood on the bridge with binoculars, and around nine o'clock he noticed a smoke trail. "Follow that smoke, Ensign Saltsman," he said. "Increase her to half throttle, but be prepared to open her up on my command. Battle stations! There's a ship ahead, and we're going to check it out."

It didn't take long at half throttle to close the distance. They were now headed almost due west with the sun at their back, seemingly an optimal approach for surprise, if they were stalking a Japanese ship.

Sherertz looked over the men and felt they were ready. He now could see the outline of the ship ahead, silhouetted against the dark sky to the west. It was easy to make out that it was a US destroyer. "She's one of ours, men!" he yelled.

He was just about to tell everyone to stand down when they saw a flash from one of the destroyer's five-inch guns and heard the shriek of an incoming shell. It hit just fifty yards in front of them, throwing up a splash fifty feet in the air.

"Evasive action!" yelled Sherertz.

"On it, Skipper," replied Ensign Saltsman, turning sharply to starboard.

It was a good thing he did, as they saw another gun flash, heard the approaching shell, and saw the water plume behind them where they had just been. The water arched up into the air and drenched the stern of their boat.

"Sparks, use the signal light and tell them who we are," yelled Sherertz.

Radioman Haas, who had been manning the forward .50-cal machine guns, fairly leaped from the gun turret and raced to the signal light mounted just in front of the bridge. He immediately flashed Morse code toward the destroyer. "Skipper, I don't think they can see our signal light with the sun behind us," he yelled.

"I think you're right," said Sherertz. "Take us north, Saltsman, full speed."

They saw another gun flash.

"Saltsman, head for the wave trough!" yelled Sherertz.

Just as they dropped into the trough between two swells, they heard the shell go right over their heads, missing them by a few yards and landing twenty yards beyond them.

"Damn, Skipper," yelled Saltsman. "I felt that one."

"You and me both," replied Sherertz, as the water from the shell impact cascaded down over the helm.

Sherertz was about to yell out another command when he saw a signal light on the destroyer and the shelling stopped.

"They finally figured out we're friendly, Skipper," said Haas.

"Stand down, everyone," said Sherertz. "Let's get back on task. Saltsman, someone on that destroyer owes us a round of drinks. Haas, send out a signal that I would like to know who's in charge over there."

Task Force 16 destroyers and Midway-based PT boats continued to patrol the waters around Midway until June 16. No Japanese ships were found, but several pilots were rescued.

Dog Days of Midway

Figure 48. Midway Officers' Club, Christmas 1942, Lt. Jack Sherertz far left., Lt. Bryant (VA), Lt. Henry S. Parker (Harvard), Lt. Chace (Yale), Ens. Bynum (Ga Tech), Ens. Saltsman. Jack Sherertz collection.

December 1942

Lieutenant j.g. Sherertz, CO of Midway Island, sat in his office on Sand reading a newspaper clipping. Hearing a knock at the door, he got up and opened it.

"Come in, Ensign," said Sherertz, without looking up.

"Why the long face, Skipper?" said the ensign.

"I just found out my uncle Lamar, a Methodist missionary in China, has been arrested by the Japs," said

Sherertz. "No one knows his whereabouts. And the worst part, there's nothing I can do to help."

"Sorry to hear that, Skipper," said the Ensign. "But right now, you have important business to attend to."

Sherertz raised an eyebrow and looked at the ensign and then beyond him at the rest of his officers standing at attention. "Follow us, sir," they said in unison, their faces attempting to be expressionless, but showing an occasional smirk. Turning as one, they marched off toward the dock.

With a grin on his face, Sherertz followed.

Down on the docks, all the enlisted men stood at ease, having assembled in several long lines facing *PT-29*. The officers turned left through a gap in the lines and marched up the gangway to the deck and forward to the bow, where they assembled facing the bridge. As Sherertz came up, they all extended their right arms to their right, indicating that he walk around them to the front of the boat where all the seamen below could see him.

Looking bemused, Sherertz did as they indicated, then stood at ease waiting for what was next.

The officers did an about-face until they were all facing Sherertz.

"Attention! Eyes forward!" yelled one officer.

All assembled, including Sherertz, snapped to full attention.

"In the absence of Lieutenant Junior Grade Sherertz's commanding officer, we the assembled officers under his command have been asked to perform a sacred duty."

With that said, all the officers lined up in front of Sherertz and pulled a piece of paper out of their pockets. Holding the paper in front of them, as though they were a choir getting ready to sing, they began speaking in unison. "On this day, December 16, 1942, we, his assembled officers, do bestow a gun deck commission upon one Jack Sherertz. Hereby having moved from the lowly rank of *one bar to two bars*, he is ordered and commanded to take upon him the duties of his exalted rank by satisfactorily wetting this evidence of his grandeur up and down."

An officer handed Sherertz his copy, and another produced a shot glass and filled it up with whiskey in front of the assembled masses.

Sherertz thought for a moment and then wet one finger in the shot glass. With dramatic flare, he ran it up and down the page. He took the shot glass and turned to face the men below. Holding the gun deck commission high in his left hand for all to see, he downed the whiskey in one gulp, then shattered the glass on the deck in front of him.

"Hip, hip, hooray!" yelled all the men. "Hip, hip, hooray! Hip, hip, hooray!"

"Thank you all," said Sherertz. "And now it's time to get down to the serious business of planning our Christmas party. We'll take all suggestions and include as many as we can. Remember, there will be a prize for winning the art contest: two extra days in Pearl on your next leave. Each entry needs to be cleared by your boat CO, and it has to be a bird. All those selected for final judging will be put on the walls of our command headquarters."

This prompted another cheer from the men.

As they walked away, the officers talked among themselves. "I'm increasingly worried about the men's morale," said Sherertz. "Everyone has seen that communiqué I posted stating that US submarines have sunk or damaged 148 Japanese ships. And we haven't even seen a Japanese ship!"

"Why don't we try coming up with some new drills?" asked another.

"I'm skeptical, but we can give it a try. We need to look into rotating the men back to Pearl for some R&R." *Maybe I'll read that news clipping about the USS New Jersey, the largest battleship ever built, that was just launched on December 7th, he thought to himself. Sixteen inch guns, my, my, my! Maybe a little distraction will help give me some ideas.*

Figure 49. Gun Deck Commission. Lt. (jg) Sherertz promoted to Lt. Sherertz.
Jack Sherertz collection.
(see page 221)

~~~

September 1943

Midway Island had been reduced to a monotonous routine. All PT boaters could fire torpedoes, strip weapons, use smoke generators, and/or perform evasive maneuvers in their sleep. They still sent out daily PT sorties looking for the enemy, but sixteen months of not a single sighting, let

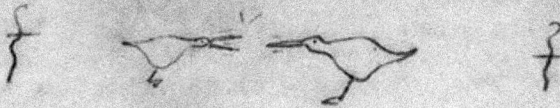

MIDWAY   Dec 16 1942

GUN DECK COMMISSION

One Jackson Shurest
hereby having moved from the
lowly ranks of "one bar" to
"two bars" (temporary) is ordered
and commanded to take upon
himself the duties of his exhausted
rank by satisfactorily writing
this evidence of his grandure
up and down.   Witness:

alone a skirmish, had taken its toll. Drinking and fighting
were increasingly frequent, and many men had been trying

to get Doc to write them out of duty for the least little thing. Something clearly was about to blow.

Today Lieutenant Sherertz was down at one end of the dock instructing some of the newer seamen in how to shoot the .50 cals. As he watched them try out what he had shown them, one of the other boat COs came up to him and spoke in a low voice.

"Sherertz, why the hell are you trying to teach that nigger cook how to shoot a gun? You know he'll just screw it up."

"First off, he's the best shot in the group. And secondly, if the Japanese attack, I would rather have him up on the deck helping with the guns than down in the galley making me a peanut butter and jelly sandwich. Try showing a little respect."

"And you call yourself a Southerner," said the CO in a nasty tone.

Sherertz was about to reply when a seaman came running up to him.

"Come quick, sir. We have a problem."

Sherertz followed the seaman back to the other end of the dock and found that a gunner's mate for his own boat had a pistol in his hands and was threatening to kill himself or anyone who came near him.

"Gunny, what's all the ruckus about?" Sherertz asked.

"My brother is dead, and I'm not doing anyone any good here, so I'm checking out today," said Gunny.

He was obviously drunk. Waving a gun in the air, he sometimes pointed it at himself, sometimes toward Sherertz and the other seamen on the dock.

Sherertz leaned over toward his XO, Saltsman. "What's going on?" he whispered.

"Yesterday he received word that his brother was killed in North Africa," said Saltsman, "and he went on a binge. An hour ago he said he was going to kill himself, and we've been trying to talk him down ever since. I think you may be the only one he'll listen to."

Sherertz headed toward the gangway. "I have to agree, Gunny. Today is a fine day to die. I think I'll join you." Sherertz took out his pistol and walked up the gangway to the boat deck.

Gunny stared. "You . . . you want to join me?"

"Sure, why not?" said Sherertz. "We've all got to go sometime. Why not today? But before we go, why don't we drink one last toast to our men? It's the least we can do."

"Thas a good ideea," slurred Gunny. "Whey's my bottle?" He turned around, staggering in the process.

Sherertz nodded to Saltsman, as he set his gun down, and the bottle of whiskey previously taken away from Gunny was produced and thrown up to Sherertz.

"I'll go first," said Sherertz.

Sherertz took a big swig, elaborately wiping his mouth, and passed the bottle over to Gunny. Gunny switched his gun to his left hand, took the bottle in his right, and put it to his mouth. As he leaned his head back to drink and started swallowing, the gun in his left hand dipped slightly.

Sherertz moved in, and in one motion, jammed the web between the thumb and index finger of his right hand in between the gun hammer and the rest of the gun, effectively rendering the gun unfireable. He clamped his left hand on Gunny's wrist so he couldn't pull the gun away.

As soon as Gunny realized what had happened, he threw down the bottle and started punching Sherertz.

"I could use a little help up here," said Sherertz, as Gunny pummeled him with his right hand.

In seconds, the rest of the crew was up on the boat and had subdued Gunny.

"Take Gunny to the brig and get Doc to look at him," said Saltsman. "Nice work, Sherertz . . . I think I'll have a drink myself. By the way, Sherertz, you're going to have a nice shiner."

~~~

October 15, 1943

Lieutenant Sherertz, CO RON 1, had been relieved of duty by the new CO, Lt. Erik Erikson. Erikson walked Sherertz to the airplane that would take him back to Honolulu and then back to the States. When they arrived at the plane, two parallel lines of men created a corridor for him to walk through. Each line was a mixture of men from RON 1 and various Marines and pilots.

"What's all this?" asked Sherertz with a smile. "Is MacArthur coming to Midway?"

"You won so much money at last night's poker game, Commodore, that it took all of us to bring it to you," said a Marine.

With that, each of the Marines and seamen brought a hand from behind his back and held out four one-dollar bills. As Sherertz walked down the line, he held open his carry-on bag, and they threw the bills in.

When he got to the end of the line, Lieutenant Erikson stopped him. "We wanted to give you a couple of going-away gifts," said Erikson, holding out several wrapped packages. "You have to open them before you can leave. The first one is from all the Marines at the base."

Lieutenant Sherertz dropped his carry-on bag and opened the first present. Inside was a pair of Marine gaiters. Upon closer scrutiny, inside one of the gaiters was the date 6-6-42, which coincided with the Battle of Midway. Then he opened the second present. It was a book entitled *Victory at Midway*, by Griffith Bailey Coale.

"Your parents sent this to you, and it finally caught up," said Erikson. "I think it should make for some good reading on the way back to the States." Erikson handed Sherertz the last package. "The last present is special, from a friend of yours who used to be here. He heard you were leaving and sent this package."

Sherertz opened the present to find a pair of practice

bombs. It came with a note that read, *I thought these might bring back a few memories. Commander Ringblom.*

"Thanks, Al," said Sherertz as he looked up toward the sky. He climbed the ladder to the plane and turned and waved.

"Attention, officer on deck," said Erikson.

The entire group turned and saluted as one. Sherertz looked as though he was about to say something, but couldn't. A tear rolled down his face, and he turned and disappeared inside.

"I think we got to him," said Erikson to the assembled men.

~~~

As Sherertz entered the plane, he wiped away the tear and noticed that there was only one other passenger. He was about to take a seat off to the side by himself to get some shut-eye, when he heard his name called.

"Lieutenant Sherertz, would you like to join me?" asked the stranger.

Although Sherertz hadn't met the man before, he thought he looked familiar, though his being dressed in civvies threw him off a bit. He walked over to meet him. "I would be happy to share the ride. My name is Jack. What's yours?"

"My name is John. Just so you're not at a disadvantage, I'm also called Lieutenant Commander John Bulkeley."

"I thought you looked familiar," said Sherertz. "You're the guy who rescued MacArthur. My parents sent me that book, *They Were Expendable*, along with a newspaper clipping. What a great story." Sherertz sat down next to Bulkeley and asked, "So what was it like, receiving the Congressional Medal of Honor?"

"Pretty stuffy. They were looking for a hero to promote the war effort and decided I was it. It took me a long time to get back to the Pacific."

"Gotta ask, John. What was MacArthur like?"

"A brilliant tactician and a royal son of a bitch."

"I can read about the tactician part. Tell me about the son-of-a-bitch part."

"I could tell you a lot of stories, but my favorite is about a motor mech second class who was in RON Two with me. A hell of a guy! As you may have read in the book, we had to leave two boats behind along the way. Both were carrying personal effects of MacArthur, including artwork. When I had to leave behind the 32 boat, because it had to dump most of its gasoline, we were able to jam everything into the remaining three boats. But after we had to drop off the 34 boat for repairs at Anaken, the two remaining boats had to absorb some generals and MacArthur's artwork. I suggested we dump his artwork to make room for the generals. He looked at me as if I were crazy and told me to get rid of some of my men. I was mad as hell, but powerless.

"My motor mech was speechless when he heard why I was ordering him off the boat. The group left behind had to fight with the Marines through some rough land battles before they were able to rejoin the Navy. Afterward, there was a ceremony for all my men for them to receive either Bronze or Silver Stars. When it came time for my motor mech to receive his Bronze Star, he refused. When he was ordered to do so, he stepped forward one step, did a perfect left face, and marched himself back to the barracks, leaving a bunch of red-faced officers screaming at him from behind."

"And what did you do to your insubordinate motor mech?"

"I bought him drinks until we both got sloshed."

"I would like to shake that man's hand," said Sherertz. "Where did they send you after you got back from the States?"

"To the southeast tip of New Guinea to command RON Seven. We patrolled the straights and islands from Milne Bay up to Vitiaz Strait."

"See much action?"

"Essentially none."

"Come on, you must have at least a few stories to tell."

"There was one weird thing that happened. You know that actor Robert Montgomery? He was in that creepy movie *Night Must Fall*."

"Yeah, I saw that movie when I was in college. It was

pretty good. He's been in all sorts of movies. What about him?"

"They promoted the XO of my command PT boat to command a boat in another squadron and announced that his replacement was arriving the same day my XO left. He stepped off the transport and introduced himself as Lieutenant Robert Montgomery. He looked pretty scruffy, with a heavy beard, but I still recognized him. When was the last time you saw an XO of a boat with the rank of lieutenant? He did a good job and didn't ask for any special favors, but it all seemed pretty strange." Bulkeley paused a moment, then said, "Say, I heard some pretty good stories about you. Some of your men told me about how you stopped a man from committing suicide and about the funeral you arranged for the Marines. That's good stuff, Sherertz. Got any other stories we can pass the time with?"

"There is one celebrity story I can tell you. Gene Tunney visited Midway as the director of fitness and athletics for the Navy. As squadron Commander , I had to orchestrate his visit. At one point after he had put everyone on Sand Island through some exercises, I suggested that we take a swim to cool off. He thought that was a great idea, right up until he saw a shark fin. The water is pretty shallow in places, and we had swum out maybe a hundred yards when we saw the first whitetip reef shark. As soon as he realized what it was, he virtually walked on water trying to get back to shore. By the time we got into the shallows, we had seen several more whitetips and a tiger shark. After that I couldn't even get him to walk near the water. I'll tell you, though, I wouldn't want to box the guy. At forty-six, he's still in amazing condition and probably could beat most amateur boxers. Everyone's got something that scares them. A bunch of sharks might scare me, too."

"Jack, I have one more question for you. Have you heard anything lately about Joe Taussig? I met him at a party thrown by his dad, and we hit it off immediately. I know he survived, but I haven't been able to find out much more."

As soon as Bulkeley mentioned Taussig and Pearl, the smile faded from Sherertz's face. It was as if someone had pulled down a curtain. Sherertz told him what he knew, but it was clear that the conversation was over.

As they began their descent onto the Ford Island runway at Pearl, Bulkeley turned to Sherertz and asked, "Have you gotten your orders yet for your next assignment?"

"No," said Sherertz. "They told me I would get them back at Melville. Meanwhile, I get to help train rookie PT boaters."

"I have a hunch we'll be seeing each other soon enough," said Bulkeley.

"I look forward to it."

On the ground, they shook hands and went their separate ways.

~~~

Wearing his dress blues and surrounded by a group of dancing girls with feather boas and flowers in their hair, Jack Sherertz spoke to the photographer preparing to take their picture. "I want you to keep snapping pictures until you've got one that will make all of my squadron back at Midway wish they were here." *Here* meant the Silver Dollar Saloon in San Francisco. "Ready, ladies?"

"Yes, Jack," said the dancing girls. "Anything you say, Jack."

"Big smiles all around," said the photographer. "Now lean in. Bend over toward him, ladies, and show me some cleavage. That's it. Now drape yourself all over him. Excellent. I've got what you need."

"There's another ten dollars in it for you if you develop the photograph by this afternoon, get it signed by the ladies, attach this note to it, and mail it today," said Sherertz.

The note read, *Lieutenant Sherertz told us all about all the handsome men stationed at Midway, so we had this picture taken so you would know we're thinking about you. If you ever get to San Francisco, come to the Silver Dollar Saloon, and we'll thank you personally for protecting our country.*

(see Figures 50 – 56 on next pages)

Figure 50. Torpedo practice. Jack Sherertz collection.

Figure 51. Smoke generation as seen from on board a PT boat (A) and from another boat (B). Jack Sherertz collection.

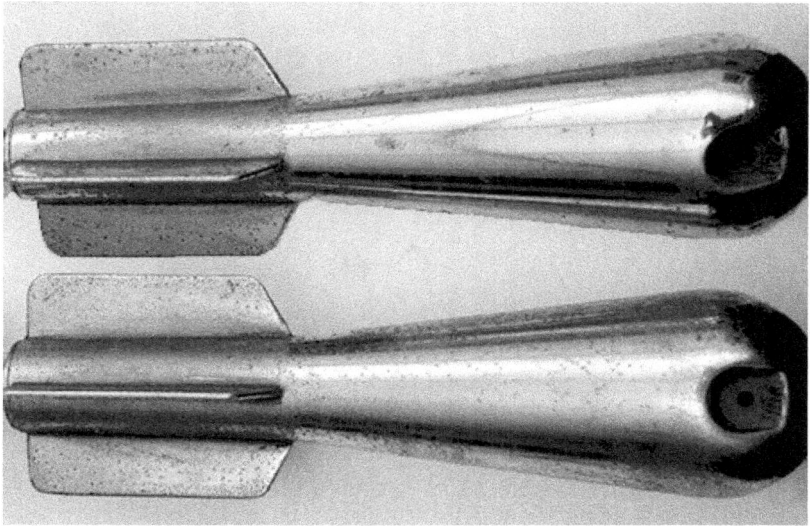

Figure 52. Pair of bombs given to Lt. Sherertz by marine aviator,
Jack Sherertz collection.

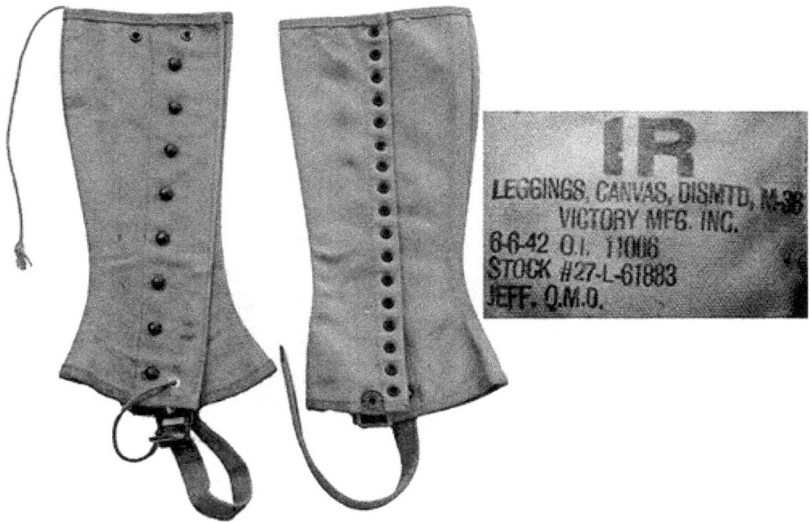

Figure 53. Gaitors given to Lt. Sherertz by Midway marines (A), date on gaitors
was June 6, 1942 (B). Jack Sherertz collection.

Figure 54. Lt. Sherertz on leave with his family in October 1943. (L-R: Frank Sherertz, Mary Sherertz, Jack Sherertz, Jeanne Sherertz, Robert Sherertz.)

Figure 55. Lt. Bulkeley Congressional Medal of Honor.
Roanoke World-News clipping, 1942.
Can be seen at www.AHeroAmongMillions.com.

Figure 56. August 19, 1943. Japanese destroyer cuts Lieutenant
Kennedy's boat in half. Roanoke World-News clipping.
Can be seen at www.AHeroAmongMillions.com.

Melville, Rhode Island

November 1943

Lieutenant Sherertz reported to the duty officer at the PT boat training center. "My name is Lieutenant Jack Sherertz. Here are my orders."

"Lieutenant Sherertz, welcome to Melville, sir," said the ensign, saluting.

"At ease," said Sherertz. He looked around while the ensign reviewed his orders.

"My name is Ensign Jacobs, sir. We really need an underway gunnery instructor. Our rookies will be very pleased to work with someone with field experience. Your first group of trainees will meet you on the dock at zero eight hundred tomorrow. Your office is down the hall on the right. Can you drop this package off at the office just before yours?"

"Sure."

"Thanks. By the way, this envelope arrived a week ago."

As he walked down the hall, Sherertz studied the envelope he was holding, trying to figure out where it came from. He found his office and knocked on the one right before it without paying attention to the name on the door, his mind still on the mail he had received.

"Come in," said a voice from within.

Sherertz opened the door, expecting to merely hand off the package. Lo and behold, there was Johnny Mote. "Well, aren't you a sight for sore eyes," said Sherertz.

"I can't believe it," said Mote. "I haven't seen you since Midway. How the hell are you, Sherertz?" as he came up and gave him a big hug.

"Really good, Johnny. The time at Midway let me recharge. How do you like instructing?"

"I like it a lot. I think you will, too. Have you heard from any of our old buds?"

"Not a one, but Midway isn't on the way to anywhere, so I wasn't expecting to."

"Hey, did you hear about Captain Bode from the Oklahoma?"

"No, what happened?"

"He killed himself. Word is he fucked up at the Battle of Savo Island, but it's all a big mystery. Say, I've got some work to finish now, but let's get a drink after work, and I'll tell you the scuttlebutt around here."

"Sounds good. Come by when you're ready," said Sherertz. He walked out and headed toward his office. Once there, he opened the envelope. It was a picture of RON 1, complete with a group of Marines, all standing together in front of the mess hall. One man in the middle held up a picture, and the rest of them were giving the photographer a salute with their middle finger. Sherertz erupted with uncontrollable laughter to the point that he quickly shut the door of his office behind him.

~~~

Lieutenant Sherertz was sitting in his office doing paperwork when there was a knock at the door. "Come in."

"I thought I would come by and introduce myself, as we'll be overlapping for a few weeks. My name is Jack Kennedy."

"Nice to meet you, Jack. My name is Jack, also. You're the guy they say is going to be a politician, right?"

"Someone must have been talking to my dad. Right now I'm just trying to get through the war in one piece."

"I know what you mean, after what I heard about your *PT- 109*."

"You must have had a pretty wild time yourself on the Nevada."

"You got that right."

"Listen, Jack, I've got a threesome for golf this Saturday. Why don't you join us, and then we can have some drinks afterward at the Officers' Club."

"What time?"

"Meet us at one o'clock in the Pro Shop."

"Great. I'll need some clubs. I haven't replaced the ones that sank with the *Nevada*."

"The Pro Shop will rent you some."

"Excellent. See you then."

"By the way, I almost forgot to give you your new orders," said Lieutenant Kennedy. He handed them over, then turned and walked out.

~~~

Sherertz walked down the hall and knocked on Mote's door.

"Come in," said Mote.

Sherertz entered and shut the door behind him. "Listen, Johnny. What's the deal with *PT-109* and Kennedy? I've been hearing all kinds of rumors, and I have to play golf with him this weekend."

"The scuttlebutt is that he thought his 109 boat was his own private speedboat and was reckless as hell. The last straw was the night his boat got cut in half by a Jap destroyer. Three other PT boats in his squadron engaged the destroyer, but the 109 boat couldn't, because Kennedy, allegedly, didn't have the engines ready to maneuver, so he couldn't avoid the collision. His CO started court-martial proceedings against him, but when they went up the ranks, FDR squashed them. It seems his Daddy, the Ambassador to England, intervened. After that the CO did the only other thing at his disposal and sent him back here. And one other thing, he's a world-class ladies' man. Another rumor going around is that he had a fling with a German spy down in Charleston just before he got into the war."

"Great! The last thing I need is another prima donna. One other thing, Johnny. It looks like I won't be teaching long. I've just got my new orders. I'll be the XO of Squadron Thirty-Four. You know anything about the CO? His name is Harris."

Figure 57. Melville RI MTB Training Center stationery logo. Jack Sherertz collection. The image can be found at www.AHeroAmongMillions.com.

Figure 58. Lt. Jack Kennedy on a PT boat. Wikipedia.

Figure 59. Melville RI PT Boat Training Center (MTBSTC), February 1945. The PT Boat Forum. The image can be found at www.ptboatforum.com/cgi-bin/ MB2/netboardr.cgi?fid=102&cid=101&tid=842&pg=81&sc=20&x=0 and at www.AHeroAmongMillions.com.

RON 34 Commissioned

Figure 60. RON 34 insignia. Jack Sherertz collection.

December 31, 1943

On December 31, 1943, Lieutenant Sherertz attended the commissioning luncheon of Motor Torpedo Boat Squadron 34 at the Lottie and Jack Restaurant, NYC. He was now the executive officer (XO) of RON 34, second in command to Lt. Allen Harris, the commanding officer (CO). Seated around the head table in the main dining room were Lieutenant Sherertz, his father Frank, his mother Mary, his sister Jeanne, his sixteen-year-old brother Bob, Lieutenant Harris and his family, and Lieutenant Junior Grade H. B. Sherwood and his family. The rest of the room was filled up with tables seating the remaining officers of the twelve boats in RON 34, with each of their families.

"I would just like to say how glad I am that my family is here," said Jack as he stood up from the table. "A toast. To family, and knowing they are always there when you need them."

Everyone at the table raised their glasses in response—everyone except one.

"Mom, can we have anything we want for dinner?" asked Bob Sherertz, totally oblivious.

"No, Bob, the menu is right there in front of you, and those are the only choices," said Jeanne.

"But I don't like chicken."

"That's enough, Bob," said Frank. "Lieutenant Harris, where is Squadron Thirty-Four heading for their first assignment?"

"We have a shakedown cruise to Miami," said Lieutenant Harris. "Then we'll go through the Panama Canal, up to San Diego, and out to Pearl Harbor. We'll get new orders when we get to Pearl. That won't be for a while. First we have to fill out the rest of our rosters and do some local training. Once PTs *498-503* are fully staffed, I'll take them up to Melville, Rhode Island, to get them used to PTs on the water. Then Jack will join me with PTs *504-509* to round things out."

"What kind of pies do they have?" asked Bob.

Everyone laughed.

"Perhaps this isn't the best time for the business of war," said Jack. "I'll tell you what, Bob. Eat whatever you like that's on this menu, and after dinner I'll take you out for apple pie and ice cream. How's that sound?"

"Now you're talking."

Figure 61. Officers of RON 34 at time of commissioning. Lt. Sherertz is located 2nd from the right in the front row. The other assembled officers include Lt. J. J. Daniel, Lt. C. E. Twadell, Lt. j.g. W. C. Godfrey, Lt. j.g. J. A. Doherty, Lt. j.g. R. C. Burleson, Lt. W. M. Ball, Lt. j.g. C. R. Whorton, Lt. j.g. R. W. Netterstrom, Lt. j.g. J. H. Kurre, Ensign B. T. Hemingway, Ensign W. S. Squire, Ensign H. G. Fraser, Ensign R. P. Cooper, Ensign W. R. DeYoung, Ensign W. D. Surgeon, Ensign R. W. Hadley, Ensign R. M. Bond, Ensign S. D. Allen, Ensign J. K. Pavlis, Ensign W. K. Sites, Ensign R. L. Youmans, Ensign J. F. Queeney, Ensign R. L. Baker, Ensign M. J. Sharkey, Ensign L. E. Pierce, Ensign W. Wotherspoon, Ensign W. E. Murphy, Ensign R. E. Schuster, and Lt. j.g. H. B. Sherwood. Jack Sherertz collection.

Figure 62. First four boats commissioned in RON 34. Jack Sherertz collection.

Shakedown Cruise

Figure 63. Inland Waterway, RON 34 shakedown cruise.
Jack Sherertz collection.

March 23, 1944

Lieutenant Harris left New York City for Miami on March 11 with RON 34 Group A (PTs *498-503*). Group B (PTs *504-509*) was now assembled at the New York City Navy dock, ready to depart for their shakedown cruise. The men were on board each of their ships.

Sherertz had asked the officers to assemble on the dock for departure instructions. "Gentlemen, it's time," he said.

"I know you all think you're ready for the war, but you're not, and the purpose of this cruise is to learn where the training gaps are and to find out if all the boats are mechanically sound. Also, I know the scuttlebutt is that this is a vacation cruise to Florida, but make no mistake about it, we are at risk of attack by U-boats whenever we're in open ocean. Hitler's U-boats have orders to sink any ship off the coast of the US, and last year they torpedoed six. We'll be traveling on open-ocean from here to Cape May, New Jersey, and from there to Norfolk, Virginia. I want each of you to deploy men with binoculars fore and aft and up top until we are safe in the Intercoastal Waterway."

"Come on, Skipper," said Lieutenant Crist. "Isn't all that stuff about U-boats and 'loose lips sink ships' just a bunch of hooey?"

"Five days ago, a U-boat torpedoed and sank the tanker *Seakay* just off our coast," said Sherertz. "It's just one of a long list of boats already sunk. We can clearly outrun them, and we have the armament to sink them, but only if we see them first. Go slow out of the harbor and then accelerate to twenty knots thereafter. Our intelligence suggests that's faster than most U-boats can travel. If you see a U-boat, you have permission to engage. Good hunting, men."

The 152-mile trip to Cape May, New Jersey, proved uneventful. At twenty knots, with one refueling, it took about ten hours. The men were clearly glad to be doing something other than drilling. As they pulled into the harbor to anchor for the night, Sherertz felt a foreboding that he couldn't initially identify. He looked at all the boats, and everything seemed in order. He looked around the harbor and only saw a few other commercial boats of no concern. Then it hit him. All the buildings in the harbor and beyond had their lights on. After seventeen months of total blackout every night at Midway, this seemed dangerously wrong. Then he remembered naval command saying that all the East Coast mayors had petitioned the president, telling him that it would be bad for tourism to have the lights out at night. *What a way to fight a war*, he thought to himself.

From Cape May they traveled 192 miles to Norfolk, Virginia. At twenty knots, with one refueling, it took about twelve hours. The ocean was rougher, and many of the men got seasick on this leg. As they pulled into the Norfolk harbor, Sherertz felt a sense of relief. They were at a naval base, and they knew a war was on. His boats would be safe here.

The next day, Group B entered the Intercoastal Waterway and traveled the two hundred miles from Norfolk to Morehead City, North Carolina, in about ten hours, not counting the two-hour stop for lunch and refueling. With dead calm water and no U-boats to worry about, this leg was a breeze. They docked at the Morehead City Naval Station about 4:00 p.m. to refuel and spend the night. While refueling and engine inspection were underway, all nonessential personnel were given leave until midnight. In short order, the men found several watering holes in Morehead City that were frequented by female Marines and settled in for dinner and a few beers.

~~~

The next AM Sherertz's officers, Aldridge, Godfrey, Brettell, Heminway, Whorton, and Netterstrom assembled on the dock in front of him and gave report. Last night's leave was uneventful except for two seamen on *PT-507* who inadvertently allowed a female marine to fall overboard off their PT, and then caused further ruckus by drying her clothes on one of the diesel engines, increasing one of their reputations as the squadron screw-up.

"I'll take the lead of Division 1 in *PT-504*, with PTs *505* and *506* right behind," said Sherertz. "Lieutenant Netterstrom, you take the lead of Division Two in *PT-509*, with PTs *507* and *508* right behind. Since everyone had such a good time last night, we had a slow start this morning and have some time to make up. Forget what I said last night about slowing to 15 knots today and continue with the 20 knots we've been running."

"Aye, aye, sir," they collectively responded, and headed to their boats.

Sherertz watched as the men boarded their respective boats. Soon, Group B was loaded up and cruising down the Inland Waterway in high spirits.

In places, the Inland Waterway looked not much larger than a big ditch. In that setting, an eighty-foot, forty-ton PT boat made a big wave. As Sherertz watched from the bridge of *PT-504*, he saw a couple standing on a dock. He waved vigorously and pointed behind their boat, trying to warn them of the coming wave. They didn't understand and gave a friendly wave back.

Sherertz turned to Ensign Sharkey, the boat XO, smiled, and shook his head. "Ensign, that couple is not going to be happy."

They watched as the wake of *PT-504* moved toward the couple. The wave hit the dock and nearly washed the pair off. The man shook his fist, but the woman just laughed and waved again. Sherertz waved, smiled, and shrugged his shoulders.

Unbeknownst to Sherertz, things were about to get exciting for *PT-506* behind them. The water pushed up on shore by *PT-504* was followed by a second wave from PT-505. By the time *PT-506* came along, the water pushed out by *PT-504* and *PT-505* had not returned to the Inland Waterway.

"The water looks real shallow up there," said Surgeon, XO of PT- 506, to his skipper, Lieutenant Bretell.

Before Bretell could respond, their boat ground to a halt in the sand. The sudden breaking effect caused the forward gunner's mate and another crew member to have to hang on to the 20 mm gun to keep from being thrown off the bow of the boat, giving the couple on the dock the last laugh.

"All stop, Surgeon," said Bretell. "Any leaking down

below?"

"No problems, Skipper," came a voice from below. "She's tight as a drum."

"Give me the mic," said Bretell to the XO. "Lieutenant Sherertz, you're washing all the water out in front of us. We just ran aground."

"Let's spread 'em out, boys," said Sherertz over the radio. "A hundred yards between each boat."

~~~

Things were uneventful after that until the PT boats got to Charleston.

"I think we'll stop for a bite to eat somewhere here in the harbor," said Sherertz into the radio. "Lieutenant Crist, you find us a place to anchor."

"Aye, aye, sir," said Crist with a big grin on his face.

Crist already had a reputation as a cowboy captain, not uncommon among men serving on PT boats. He had his XO gun their boat and took the lead. *PT-509* did two back-to-back power turns, which made the channel bottom difficult to see for *PT-506* following immediately behind. As a result, *PT-506* waited too long to turn and ran aground on a sand bar. The seaman who had been standing next to the forward gunner grabbed for the gun a second time to keep from being thrown off the boat.

"Damn that Crist," said the seaman who almost took a bath.

PT-504 pulled up next to *PT-509* in time for Sherertz to see that *506* had run aground. It was clear that Sherertz was not happy.

"This is a good place to stop," he said. "We'll break for an hour and then cruise until dark. Lieutenant Crist, your showboating led to *506* running aground. Part of the purpose of a shakedown cruise is for everyone to learn what to do and what not to do. If this was a battleship, someone would get court-martialed. If the bottom was rocky, we would have lost *506*. I do not want to see this happen again. Are we clear?"

"Aye, aye, sir," said Crist.

The rest of the men went swimming or played baseball on the sandbar. Crist was briefly subdued after his public chastising, but soon his irrepressible personality returned.

~~~

April 9, Miami, Third Naval District Training Center

Group B had been out drilling all day and returned in the late afternoon. Group A was already there on the dock waiting, having finished their shakedown exercises. Both groups had been instructed to shower up and assemble for an important announcement. Soon the men were huddled together in several groups near the dock where the boats were moored. It was 96 degrees, with greater than 90 percent humidity. By the time Lieutenant Sherertz walked up, it looked like they had all forgotten to take a shower.

"Lieutenant Sherertz, want to get in on the action?"

asked Ensign Schuster.

"What kind of action?"

"The pool! We're all betting on when we'll arrive at Pearl. Check it out." Schuster pointed to several pieces of paper pinned to the board where daily orders were posted. Every day between now and July 1 was listed. It was clear that the greatest interest lay in a one-week period in mid-May. Multiple names had been written after each day of the week, and next to each name was a time of day so that the closest to the arrival time could be determined. Sherertz was studying the options when Lt. Harris came up behind the group unnoticed.

"Attention! About face! At ease! I'm impressed," said Lieutenant Harris. "You've picked the right week, but the wrong ocean."

"Wrong ocean, sir?" said Ensign Allen.

"New orders. We're going to Plymouth, England. We have two weeks to get back to New York City so they can add new armament to our boats and paint them gray instead of the current Pacific green. I'll be flying over before you so I can get things ready in advance. Because of the German U-boats, they're splitting up our squadron into five groups that will be transported on five different ships. Each boat will be accompanied by a skeleton crew of one officer and five enlisted men. Most of you men will ride over with Lieutenant Sherertz on the HMS *Oceanway*, leaving on May second. A few will leave earlier on April eighteenth to help me get our new base ready in advance. Any questions?"

"What's the gig, Skipper?" asked Ensign Russ Schuster.

"We'll get our orders when we get to Plymouth," said Harris. "Group A leaves tomorrow, Group B on the nineteenth, after completing their shakedown exercises. Dismissed."

As they broke up, Ensign Heminway came up to Lieutenant Sherertz. "Lieutenant, can I speak to you a moment about my gunner's mate, Bos Bosley?" he asked.

"What's on your mind, Ensign?"

"He's driving me crazy, sir," said Heminway, looking quite exasperated. "First he used an hour's worth of fuel to dry a lady Marine's clothes. Then he refused to clean out the bilge water when ordered by a chief petty officer, until I agreed with the order. And now he's in the hospital with an infected finger. The doc says he needs to stay a few more weeks for it to heal, but Bosley says he'll leave against medical advice. He's not going to miss the war because of a little infection. Every time I turn around, he's saying, 'You got to get me out of this, Skipper.' What should I do?"

"So he's the one," said Sherertz, with a grin. "How old is Bosley?"

"Seventeen."

Sherertz smiled. "Do you think he'll follow you into battle?"

"I think he'll follow me anywhere."

"Is he any good as a gunner?"

"He's the best I've got, and he's not afraid of anything or anybody."

"Sounds like a hell of a sailor to me and just what we need to win the war," said Sherertz. "If I were you, I'd get him out of the hospital and take him with you. Where you going, Ensign?"

"I'm going to get my gunner's mate out of the hospital so I can take him to England," said Heminway over his shoulder, as he walked away with a big smile on his face.

Figure 64. RON 34 PT boats, Inland Waterway shakedown cruise.
Jack Sherertz collection.

HMS Oceanway

Figure 65. HMS Oceanway, used to transport RON 34 boats and personnel to England. Wikipedia.

May 3, 1944, 10:00 p.m.

RON 34 officers were gathered in the officers' mess on board HMS *Oceanway* when in walked Lieutenant Sherertz, taking the seat at the head of the table.

"What's the story, Skipper?" asked Ensign Schuster.

"We've cleared New York City harbor, as you know, and we're heading to take our position in Convoy CU

Twenty-Three."

"How big's the convoy?" asked Ensign Allen.

"Forty-nine ships with 39,288 troops, escorted by 12 destroyer escorts and a light cruiser.

Lieutenant Crist whistled loudly.

"Cruising speed will be fourteen knots with an estimated arrival time in Liverpool, England of May 14."

"This should be a real snoozer," said Crist. "Can't you speed them up? I want to get to Europe before the war is over. Besides, I hear the limeys have sunk the German navy, so we don't need escorts any more." He had a cocky look on his face, as opposed to his fellow officers, who looked nervous.

"First of all, Lieutenant Crist, CU convoys are fast convoys; most convoys only move at 8 knots," said Sherertz. "Secondly, don't be in too big a hurry to fight the Germans. War isn't like what you read in dime novels. Before you get too comfortable snoozing on your deck chair, sipping mint juleps, you should be aware that earlier today, destroyer escort USS *Donnell*, part of Convoy CU 22 a few days in front of us, was sunk by a U-boat. Twenty-nine killed and twenty-five wounded." *That got the attention of all the men, thought Sherertz as he looked around, except Crist. He still looked impatient. I think I need to keep this one busy.*

"All right Lieutenant Crist, to make sure this convoy is not a snoozer for you, I have a job for you. I want you to organize the men into a round the clock observers group.

You will be looking for submarines and to tell if the ships in our part of the convoy stay in formation."

"How will I know if a ship is out of position," said Crist.

"Each ship will be in a column 5-6 ships wide, with 600 yards between each ship," said Sherertz. "Each column will be 1000 yards behind the column in front and 1000 yards ahead of the column behind, with nine columns in total so that the convoy will fit in a box roughly 3000 yards wide by 8000 yards long. The destroyer escorts and cruiser will cruise in an oval ring around the convoy – just like we used at Pearl to protect battleships and carriers. If you see big gaps develop, let me know immediately. Is that clear?" As he spoke he looked at Crist and thought, *if all my men were like Bosley and Crist, this war would be over in a month.*

"Aye aye, sir."

"Anyone else have any questions?" said Sherertz.

"What happened to the U-boat that sank the destroyer escort?" asked Allen.

"The Second Support Group under Captain Frederic Walker hunted it down, brought it to the surface with depth charges, and sank it with gunfire. So keep your eyes open and let the bridge know if you see anything that looks like a periscope. Are we clear?"

"Aye, aye, sir," said the assembled officers.

"Have any of you found volunteers for the positions of squadron historian or squadron photographer?" asked Sherertz.

"Bosley from my boat agreed to be the historian," said Ensign Heminway.

"Several men from different boats have cameras and are willing to contribute photographs," said Lieutenant Whorton.

"Where did you get the idea for the historian and photographer?" asked Lieutenant Bretell.

"Let's just say someone in the film industry started me thinking," said Sherertz. "Anything else?"

"Skipper, what was it like at Pearl Harbor?" asked Allen.

Sherertz looked around at the group of officers and shook his head. "To quote a friend of mine, it was the goddamn biggest FUBAR I've ever been in. They caught us asleep at the wheel."

"But the *Nevada* must have gotten in a few good licks," said Crist.

"Sure, we got a few of them," said Sherertz.

"Give us details," said Crist.

Sherertz's face took on a dark look.

Ensign Schuster had been shaking his head no, but Crist didn't see him.

"You want details. I'll give you details. In a matter of minutes, hundreds of people that I knew were killed. Some were blown into little pieces. Some were never found. Some were buried alive. Is that enough details?" Without another word, Sherertz got up from the table and walked out of the room.

"Nice work, Crist," said Schuster.

"Why's he so ramrod stiff about Pearl Harbor?" asked Crist.

"I guess we'll know once we've been in battle," said Allen.

Weymouth, England

May 16, 1944

The HMS *Oceanway* arrived at Liverpool, England and unloaded the PT boats and PT boaters. Sherertz had each crew check the fuel and the engines and then head out of the harbor. They turned south past Wales and then east, after rounding the southwest tip of England. They hugged the coast until they reached Weymouth harbor. There they tied up at the Plymouth naval dock adjacent to Portland Bill, a large rock peninsula that protected the seaward side of Weymouth Harbor. Weymouth had been a town of about thirty thousand before the war, but it recently had exploded with US and British soldiers and sailors readying for the D-Day invasion.

"Lieutenant Sherertz, nice you could make it," said Lieutenant Harris.

"What's the deal, Skipper?" asked Sherertz.

"The men and boat officers will remain billeted on the boats and eat at Tent City, a mile up the Portland Bill," said Harris. "You and I will stay in Tent City, and buses will take us back and forth. Some of our boats will be arriving here, some in Liverpool, and some will end up in Scotland. I'll let you know as soon as we have more details so you can round them up. Now I would like you to meet someone."

Sherertz turned around, and there stood the imposing figure of Commander John D. Bulkeley. He had on a navy blue jacket and a baseball cap that was tilted well to one side.

Before Harris could say anything else, Bulkeley came up and shook Jack's hand. "Good to see you again, Sherertz. Let's get a move on. We've got a lot to do."

With his arm around Sherertz's shoulder, Bulkeley pulled Sherertz away at a brisk pace, with Harris hustling to catch up.

Sherertz called back over his shoulder to the group of officers looking toward him for orders. "Get everything unloaded from the *Oceanway*, and I'll be back as soon as I can."

~~~

Bulkeley, Harris, and Sherertz arrived in front of Tent City.

"I hate to do this to you, Sherertz, but I'm going to reunite you with an old friend of yours," said Bulkeley. "Follow me."

Bulkeley headed into a tent, with Harris and Sherertz following. Bulkeley and Harris stepped aside to reveal Commander John Ford.

"I understand you two know each other, so you'll have no problems being tentmates," said Bulkeley.

"It's an honor to see you again, Lieutenant Sherertz," said Ford. "I look forward to our continued relationship."

Sherertz looked beseechingly at Bulkeley and Harris but got no comfort. "We'll make the best of it, sir," he said with a sigh.

"Call me John, and I'll call you Jack," said Ford. "Commander Bulkeley, can I interest you in some fine Kentucky bourbon?" Ford reached behind him and opened a chest full of liquor bottles.

"Perhaps another time, John," said Bulkeley. "Jack, Ike wants Ford to ride with us and make some movies of our boys and anything else exciting that he happens to see. We'll see you later, John." He winked at Sherertz.

Bulkeley, Harris, and Sherertz exited the tent together. Bulkeley and Harris turned and smirked at Sherertz.

"You're having entirely too much fun with this," said Sherertz.

"Who knows—hanging out with a Hollywood movie mogul may have perks," said Bulkeley. "Rest up today Sherertz. Meet me in the officer's mess at 0700. There's a lot of new stuff I want you to go over with the men during the next few weeks."

~~~

May 28, 1944

RON 34 was now fully operational and out on maneuvers. Sherertz addressed the squadron over the radio.

"I have a special treat for all of you today. Follow our boat. I'm going to introduce you to an American classic."

~~~

RON 34 slowed down coming out of a light fog, and there in front of them was the USS *Nevada*, fully refitted.

"The captain is going to give me a brief tour of the boat," said Sherertz. "While I'm gone, anyone who cares to can go on deck and walk around."

Lieutenant Sherertz's PT boat pulled up to the gangway, and he disembarked and walked up. When he arrived on deck, he was greeted with a salute from the captain of the *Nevada*. To the captain's right stood a double row of seamen, at ease.

"USS *Nevada* officer from Pearl Harbor on deck," said Capt. P. M. Rhea. "Attention. Ready for inspection, Lieutenant Sherertz."

"This isn't necessary, Captain," said Sherertz.

"You've earned it, Lieutenant."

The captain led, and Sherertz followed down the line. When Sherertz finished his inspection, the captain gave him a quick tour.

"I like what you've done with my boat," said Sherertz with a grin. "New deck, new paint. We sure could have used all those antiaircraft guns when the Japs attacked us at Pearl."

"You should see them track an airplane," said Rhea.

"Let the Luftwaffe come. We're ready."

They had now come full circle back to the gangway.

"Wait here a moment," said Rhea. "I have a surprise for your men."

~~~

The PT boats clustered around the gangway at the base of the *Nevada*, ready to go.

"What's keeping Ramrod?" asked Allen. "I want to get back to base."

"There's the XO now," said a seaman.

They turned to see Sherertz coming down the gangway, followed by a line of seamen, each carrying a cardboard cylinder.

"What's this all about, Skipper?" Allen asked.

"Anyone want some ice cream?" said Sherertz

After weeks of training on board a PT, without refrigeration, ice cream was a gift from the gods. A collective cheer erupted.

Figure 66. PT boat dock at Portland Bill, England.
Jack Sherertz collection.

Figure 67. Weymouth, England, 1944. Tent City below.
Jack Sherertz collection.

Figure 68. Tent City at Weymouth, England, PT base, where officers slept. Jack Sherertz collection.

Figure 69. USS *Nevada* refitted after Pearl Harbor, September 1944. US Navy National Archives (Photo # 80-G-282709), Wikipedia.

Comdr Ford and Other Final Details

June 3, 1944, dawn

Bulkeley and Sherertz stood on a dock near Portland Bill at first light, scanning the harbor. It was full of naval vessels, mainly US, but with a large contingent of British vessels and a few Australian ones as well. Most were transport ships, but scattered among these were destroyers, destroyer escorts, and farther out, some cruisers. The battleships were elsewhere. A steady stream of small craft brought troops, vehicles, and supplies from all over southeast England to the transport ships.

Taking one final look Bulkeley and Sherertz walked off the dock and got into a jeep.

"This is going to be a long day, Sherertz," said Bulkeley.

"It's finally going to happen, isn't it?" Sherertz asked.

Bulkeley smiled and then said, "Let's head up to London and have a drink with your buddy, Ford."

You'd have to be deaf, dumb, and blind not to be aware that the invasion of Europe is about to begin, thought Sherertz.

Bulkeley began their trip by driving on Portland Beach Road, which bordered the harbor. He had to pick his way between trucks carrying US troops and British troops on foot. It was an awe-inspiring sight.

"Too bad they won't let us take any pictures," said Sherertz.

Bulkeley nodded in agreement.

Sherertz and Bulkeley continued slowly north through the streets of Weymouth around southbound military traffic. Once clear of Weymouth their plan was to head slightly northeast until they got to the air base at Warmwell, then east toward the air base at Middle Wallop, and finally northeast again until they got to London. High command had said this route should avoid most of the military traffic heading toward Southampton and Portsmouth.

Some of the men they passed were in trucks or tanks, but most were on foot. Each man carried a fully loaded pack, a gun, a canteen, a helmet, and an M1 Garand rifle — most with condoms tied over the barrels. The usual banter that came from soldiers was largely missing, and the absence of the jawing and the insults produced a feeling of tension in the air that was almost palpable. Sherertz and Bulkeley shared a look that said they looked forward to getting beyond all this converging traffic.

"Sherertz, can you believe over a thousand men bivouacked here only a few days ago?" asked Bulkeley as he pointed to their right.

"They obviously left in a hurry," said Sherertz, with a growing sense of foreboding.

Each encampment or "sausage" looked as if it had been

hit by a tornado, with trash everywhere—churned-up earth, oil stains, empty gas cans, and spent rounds of ammunition - no longer covered by camouflage nets.

"Pretty weird," said Sherertz. "Total abandonment with a sense of urgency. It makes my hair stand on end."

He and Bulkeley couldn't help notice the many civilians standing along the side of the road, smiling and waving to them, many with tears running down their face. If they came to a stop someone would come running up, frequently a child, offering them food or a drink.

They think we're going to save them, thought Sherertz.

"I think this is only the beginning," said Bulkeley. "Let's get to where we're going."

~~~

Torpedoman Easterly from RON 34 slipped under the camouflage net of one of the few remaining unevacuated sausages, immediately adjacent to Weymouth, to avoid the MPs at the main entrance. All the men there were dressed in battle fatigues with their packs lined up on the ground, ready to move out. They only awaited an order from their CO. Meanwhile, a poker game was in process to kill time.

"Nice you could make it, Navy puke," said a soldier with a private stripe on his arm. "Did you bring your money? Most of us are running a little lean and are looking for a cash infusion before we ship out. Ante up, swabbie, It's five bucks a hand."

"Try and take it, sausage grunt," said Easterly. "Last

time, I believe I left with your money in my pocket. And by the way, how's your jaw?"

"About the same as your lip, but I don't have time to punch you out today, so I'll settle for your money."

"You said it, man. It's about time we got this show on the road," said another soldier with sergeant stripes. "I'm tired of all this waiting. If they show us one more movie, I'm going to throw up."

"Amen," said many of the men simultaneously.

"By the way, PT man, did you bring any of your home brew?" the private asked hopefully.

"Of course!" said Easterly. "My canteen is full. Being that this is kind of a special day, I've decided to forego my usual charge. Drinks are on me."

"You are the man," said the private. They all held out cups, and Easterly poured each of them a drink.

"I propose a toast," he said. "To kicking some Kraut ass!"

"Hoorah," said all the soldiers.

"Besides, once our squadron moves into the Channel, I'll have to shut down the still for a while," said Easterly. "Say, did you guys get hit by any of those bombs last night?"

"Nah, those Krauts couldn't hit the broad side of a barn," said the sergeant.

A lieutenant poked his head into their area. "Saddle up,

we're moving out."

"That's my cue, boys," said Easterly as he picked up the camouflage netting and stepped under it. "Good hunting on the beach."

"Sink a ship for me," said the private as Easterly exited, and they started picking up their packs.

~~~

It was now dark, and Bulkeley and Sherertz were nearing London in total blackout.

"Skipper, got any more MacArthur stories?" Sherertz asked.

"Perhaps the most astonishing thing to me was that he was able to talk brilliantly and have a command presence, in spite of being almost continuously seasick. Sometimes he would stop midsentence to lean over the side and throw up and then pick up the sentence right where he left off. An amazing man."

"John, you're driving a bit fast for blackout conditions."

"Jack, I have eyes like an owl and sonar like a bat. You're in good hands."

"Somehow I feel safer with you on a PT boat. By the way, why are we driving at suicide speeds?"

"As you have deduced, our wait is over. We're about to get our orders, and then we'll have to drive back to Weymouth tonight."

~~~

Bulkeley and Sherertz arrived at Claridge's, a hotel in London.

"Sherertz, you get her parked and go inside and get us a table for dinner. I'll go up and check on Ford and be down in a few minutes."

Bulkeley ran up the stairs and down a hallway until he reached room 223. He knocked and got no response. He knocked again, louder this time.

The door opened and there stood Commander John Ford, naked, saluting him.

Bulkeley shook his head and grinned.

"Bulkeley, great to see you," said Ford. "Come on in, and we can talk about the movie."

"Sherertz is downstairs waiting," said Bulkeley. "Why don't you get dressed, and we'll talk over a meal."

"I'll get dressed, and the girls and I will be down in ten minutes."

"See you in the restaurant, John," said Bulkeley as he turned and left.

~~~

Ford, Bulkeley, Sherertz, and two starlets shared a table in the hotel restaurant.

"John, Hollywood wants me to make a movie about your rescue of MacArthur," Ford said. "A script is almost done. They're calling it *They Were Expendable*."

"Just like the book," chimed in Sherertz. It was the story

of Bulkeley's PT squadron's rescue of MacArthur from the Philippines.

"John Wayne has agreed to be in the movie, and Robert Montgomery has agreed to play you," said Ford, addressing Bulkeley. "But I'm not ready to sign on yet. I need some experience on board a PT. Can you help me?"

Bulkeley sighed and rolled his eyes. He looked across the table at Sherertz for help.

Sherertz ignored him as he was having a great time seated between the two starlets.

"Look, John," said Bulkeley. "Are you serious about this?"

"As serious as a heart attack!" said Ford.

"All right, John," said Bulkeley. "Here's my offer. When there's going to be some action, I'll take you with me, as long as you can stay out of the way. If one time you make me sorry, Lieutenant Sherertz or I will personally push you over the side of the boat." Bulkeley winked at Sherertz.

"Excellent. You just say the word, and I'll be ready to go in ten minutes. Now that we have an agreement, let's have an early dinner."

"John, how about if we do a little dancing first?" said Sherertz.

"Sorry, Jack. They're with me."

"We don't have time, anyway, Jack," said Bulkeley. "As you will recall, we have another appointment."

"Sorry, girls, I'm needed elsewhere," said Sherertz with a grin.

The two starlets made a show of looking disappointed and simultaneously kissed him on both cheeks. With a big smile on his face, Sherertz turned and tipped his hat to them, and then he and Bulkeley departed.

D-Day Orders

Bulkeley and Sherertz arrived at Southwick House, the Supreme Headquarters Allied Expeditionary Force (SHAEF), previously the R.N. Navigational School with the name H.M.S. Dryad, located a few miles from the British naval base at Portsmouth. They got out of the car, and showed their credentials. A guard led them to a room with another guard at the door.

After knocking and sticking his head inside, the second guard turned around and faced Bulkeley and Sherertz. "You may go inside now."

As they entered the room, Rear Adm. Alan G. Kirk, US Navy, D-Day Commander of the Western Forces (Utah Beach, Omaha Beach) and Rear Adm. Sir Phillip L. Vian, British Royal Navy, D-Day Commander of the Eastern Forces (Gold Beach, Juno Beach, Sword Beach), were in the middle of a heated discussion. Kirk turned toward them and motioned for them to sit down.

"I tell you, Phillip, our boys will give us the greatest chance of success," said Kirk.

"Those bloody wooden bathtubs of yours will be blown out of the water, and the whole invasion will collapse," said Vian.

"If German E-boats attack, they'll run circles around your ships, and before you figure out how to shoot them, they'll sink you."

"Your boats certainly are faster than our destroyers or our PTs, for that matter, but they don't have enough fire power."

"Pound for pound they are the best-armed boats in our Navy," said Kirk. "And besides, Phillip, your destroyers are so damned noisy the Germans will hear you coming miles off."

Bulkeley and Sherertz looked back and forth at each man, following the conversation.

"And one more thing: they just got here," said Vian. "They can't possibly be ready in time."

"Of course they'll be ready," said Kirk.

Bulkeley cleared his throat. "I guarantee all of our boats are ready now or at any time you need them," he said. "Don't you agree, Lieutenant Sherertz?"

"All our PTs at Portland Bill, are stocked, staffed, and seaworthy," said Sherertz. "Each boat carries twin fifties, a forty-millimeter antiaircraft gun, a twenty-millimeter cannon, two torpedoes, and depth charges. It will be plenty for whatever you want us to do." He paused. "May I ask what it is that you want us to do?"

Both admirals laughed and turned toward Bulkeley and Sherertz.

"Well, Bulkeley," said Kirk, "I guess you kept your

word on secrecy. Lieutenant Sherertz, we're talking about D-Day, the invasion of Europe. RON 34 will escort our minesweepers to clear the Channel for all ships in Force U and then take up positions along the Mason Line to guard the western flank of the invasion forces. Can RON 34 do that?"

"Yes, sir," said Sherertz. "No problem, sir."

Admiral Vian looked first at Admiral Kirk and then at Commander Bulkeley. Both nodded.

"All right, Lieutenant, I'll come to your dock later today," said Vian. "If everything is shipshape, you'll rendezvous with the sweepers at twenty-three hundred hours tomorrow, with the mission to commence at midnight. The area to be cleared is shown on the charts on the table in front of you."

Sherertz and Bulkeley looked over the charts.

"I see no problem with any of this," said Bulkeley. "Do you agree, Lieutenant Sherertz?"

"Yes, sir," said Sherertz. "One advantage of going first is that there's nothing to run into, with the possible exception of mines." He grinned.

"Lieutenant Sherertz, I know you've been at Pearl Harbor and Midway, but this is the big one," said Kirk. "Make us proud."

"Godspeed," said Vian.

"Have you heard any update on the weather," asked Kirk.

"Only that our assembled braintrust can't agree on anything, and Captain Stagg's not hiding it very well," said Vian.

"Years of planning, and the greatest invasion in the history of mankind stands at the mercy of weathermen," said Kirk. "I think Ike's about to have a stroke."

"Wouldn't surprise me a bit, with his four packs of Camels per day and who knows how many cups of coffee," said Vian.

Clearly dismissed, Bulkeley and Sherertz exited the room. As they closed the door and walked away, Sherertz spoke to the guard. "Do you have a secure line?"

"Yes, sir. Two doors down on the right."

Sherertz got on the phone. "Lieutenant Daniel, this is XO Sherertz. I need you to roust everyone and get all of our boats fully armed, fueled, and ready for inspection by thirteen hundred hours today. We have a mission. I'll explain when we get back. Do you understand? ... Excellent. I'll be back in a few hours."

Sherertz nodded to Bulkeley, who looked satisfied, and they exited the room.

Figure 70. Supreme Headquarters Allied Expeditionary Force (SHAEF) in Southwick House, a few miles from Portsmouth Naval Base. Wikipedia.

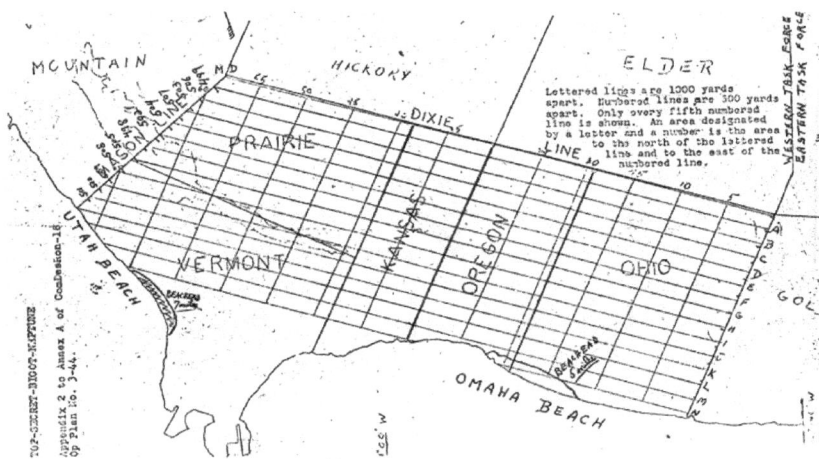

Figure 71. Operation Neptune D-Day RON 34 PT Boat Locations on Mason Line. Starting from shore they were located every 500 yards as follows: 501, *500, 509, 508, 505, 498, 502, 504, 507, 503, 506, 499.* Jack Sherertz collection.

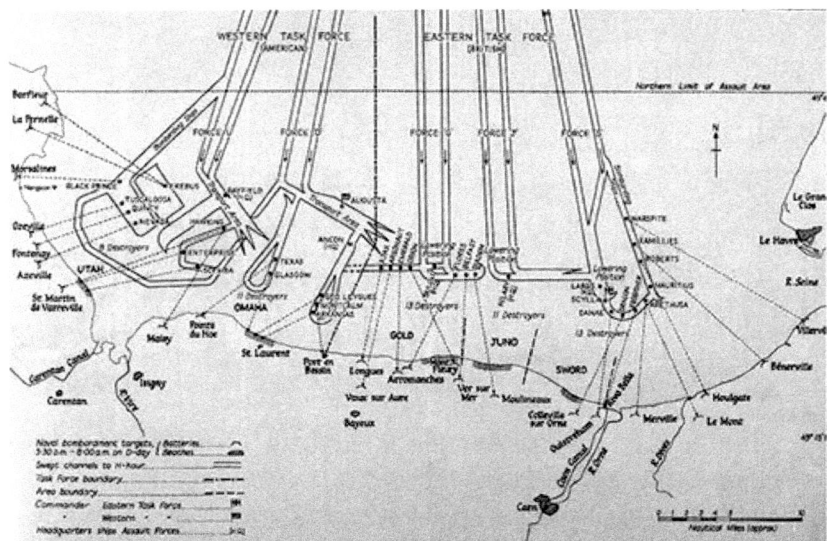

Figure 72. D-Day Naval Bombardment Targets. Wikipedia.

D-Day Rescheduled

Figure 73. HMS *Saumarez*, British destroyer. Wikipedia.

June 4, 1944

RON 34 left Weymouth, England, heading east, to rendezvous with minesweepers due south of the Isle of Wight

"Skipper, how will we rendezvous with the sweeps?" Lieutenant Kennedy said to Sherertz.

"We'll meet them just north of the mouth of the Force U invasion channels. Each channel to be swept, in our case channels 1 and 2, should have two H.D.M.L.'s (harbor defense motor launch) located at each side of the northern entrance to the channel," said Sherertz. "The H.D.M.L. personnel were specially trained to detect sonic buoys dropped at D minus 6. Each buoy has a timer that will not activate until D minus 1. Once the buoy is located the H.D.M.L.'s will line up east and west of the buoy to serve as markers to the channel entrance."

They met the sweeps uneventfully, entered the invasion channels and headed southeast flanking the sweeps. The wind, rain, and ten-foot swells made the going slow, but the group was moving forward as planned. Here and there they encountered pockets of fog. They had gone about forty miles on this southeasterly course when the radarman signaled that there was something ahead.

"Dead quiet," said Sherertz. "Full slow. Everyone stand ready."

They inched forward and became aware of a ship's engine ahead.

"Full stop," said Sherertz in a very soft voice.

As they waited to see if the ship would pass, or if they would have to open fire, they realized it was slowly coming closer.

"Signalman, give them the coded challenge," said Sherertz.

The signalman went to the blinker light and sent the message. There was no response.

"Signalman, do it again," said Sherertz.

He sent the message again, this time a little slower and

more deliberate. Again, there was no response, and again the engine sounds told them the ship was coming closer.

"Signalman, do it one last time," said Sherertz. "Everyone, listen up. If there is no response this time, be prepared on my command to open fire with all guns and then fire a torpedo and run."

The message was sent. No response occurred. Sherertz had steeled himself to give the command to open fire when the outline of a ship appeared and they saw a flag unfurl over the side. It was a British Union Jack.

"It's a limey destroyer, Skipper," said the signalman. "They've been looking for us. They're moving closer so they can talk."

Although visibility was quite limited, *PT-500* could soon make out the outlines of a destroyer about twenty yards away. Then they heard a voice through a megaphone.

"Bloody bad weather for an invasion," said a British sailor. "Eisenhower and our blokes have called it off until tomorrow night. Sorry we couldn't reach you sooner. Cheerio!"

"Can you see the name of that destroyer?" Sherertz called out to Lieutenant Kennedy, CO *PT-500*, standing next to him on the bridge, looking at the ship through binoculars.

"HMS Saumarez," said Kennedy.

"That's a wrap for the night," said Sherertz. "Let's head back to Weymouth. Whoever's in charge of that boat is going to get a bottle of scotch from RON 34 as soon as I catch up with him."

~~~

When RON 34 arrived at the Portland Bill dock, an armed garrison met them. The captain came up to the boats.

"Who's in charge here?"

"I am. I'm Lieutenant Sherertz. What's the problem?"

"No problem, sir. I just can't let you disembark. You'll be quarantined here until tonight."

It was immediately evident to Sherertz what was going on. They couldn't afford taking the chance of someone inadvertently telling the rest of the camp about the invasion for fear of a spy learning of the attack.

"Thank you, sir," he said. "We'll be fine on the boats."

"One more thing," said the garrison captain. "There's someone who wants to meet you."

A British captain came on board. "Lieutenant Sherertz, I presume. My name is Peter Cazalet. Bloody close squeak. That was my destroyer, the Saumarez, you almost blew out of the water. Apparently we just couldn't see your blinker challenge because of fog. Thanks for holding back." He held out his hand and Sherertz shook it.

"Thanks for finding us, at the risk of your boat being blown up by unswept mines," said Sherertz. "It would have been the biggest FUBAR of the whole war if you hadn't."

"When things settle a bit, I'll have you and your officers on board my ship for drinks."

Cazalet turned and left without another word.

"Lieutenant Daniel, I want you to spread the word to the other officers on board," said Sherertz. "I'm going to the other boats to explain to their officers, and then I'm going below to close my eyes for a few minutes before sunrise."

The officers told the men, and the waiting began. Some of the men went below to sleep or to listen to Armed Forces Radio.

When Sherertz came back up on deck an hour later, it was just before dawn.

"Why don't you get some rack, Lieutenant Daniel," said Sherertz.

"Don't mind if I do," said Daniel.

"Where's Bulkeley?" Sherertz asked.

"Catching a few winks on the bow of the boat. I'll see you later."

Sherertz went forward. There was Bulkeley, lying on the deck with his head on some coiled rope, with a blanket over top, the way he slept every night on the water. He had told one of the men that he never slept below, because if a torpedo hit, he wanted to be where he would be blown overboard, but still be alive.

Bulkeley opened his eyes and looked around. Seeing

Sherertz, he got to his feet and started toward him. "Are you ready, Sherertz? We've had our dress rehearsal. I'm confident it's going to go off today without a hitch."

"I'm ready, Skipper." Sherertz felt some anxiety, to be sure, but his men had done well last night, so he felt sure they would do so again when the action started.

They both looked out across the harbor. Most of the men had stayed on deck all night. With sunrise, Sherertz could see that the harbor looked like a parking lot.

Let's get this party started, thought Sherertz.

D-Day

The twelve PT boats of RON 34 headed toward their second rendezvous with the minesweepers of Mine Squadron Seven, Force U, south of the Isle of Wight. As they did, they passed a battleship at anchor.

"Skipper, I think you better check this out," said a seaman. "Isn't that one of your training buggies, the USS *Arkansas*?"

"You bet it is," said Sherertz. "I feel like I'm at a battleship reunion. Back to work, gentlemen."

~~~

RON 34 met up with the minesweepers uneventfully, just as they had the night before. There were twelve sweeps in this grouping, divided into groups of three sweeps in column formation, each column three hundred yards lateral to the next. Sherertz and Bulkeley were on *PT-501*, following the port column, immediately adjacent to the USS *Osprey*, with the USS *Auk* in front and the USS *Raven* in the rear. Eleven other PTs were spread out behind the other sweeps.

Sherertz stood on the bridge next to the ship's CO. He allowed himself a brief moment to look around. Soon boats would be arriving from Torquay, Brixham, Dartmouth, and Plymouth, to follow them down channels one and two to Normandy. He glanced upward, searching for the planes he could hear but not see. He heard movement to the left and turned just in time to see the back of Bulkeley as he left the bridge. "I'll be back in a few minutes, as soon as I'm sure everything's shipshape," he called softly over his shoulder.

"Skipper, I think the minesweepers just slowed to sweep speed," said Whorton. "They must have found the H.D.M.L.s."

"We'll have confirmation in a moment," said Sherertz.

Sure enough the sweeps had begun dropping lighted Dan buoys to show the limits of the swept water. Each lane of Dan buoys was color specific. Force U's were red and green.

As they watched the 501 cook came up from below handing out sandwiches to all on deck, no easy task given the chop they were traveling through. Task completed, he climbed back down the ladder and Sherertz heard, "It's beginning to look a lot like Christmas."

Sherertz immediately hissed to the boat CO, "Get down there and tell the men to pipe down, or I will personally throw them overboard." No further noise was heard from below deck.

Figure 74. USS *Osprey* Minesweeper. US Navy National Archives, history.navy.mil, Photo # 19-N-23989.Wikipedia.

Just before 1800 hours, Sherertz was watching the minesweeper USS *Osprey* off their port side. Suddenly there was a loud explosion, and the *Osprey* rose violently into the air. The force of the explosion flung one man inside the cabin up against the roof like a rag doll, silhouetted against the fire that had erupted behind him, and blew five or six men on the bow overboard.

"Skipper, the *Osprey* hit a mine and is on fire," said Sherertz to Bulkeley, who had just come on deck from the charting area below. "There are men overboard aft of the *Osprey*." As they watched, the *Osprey* veered in their direction.

"Watch out, XO," said Sherertz. "Permission to pick up survivors, Skipper?"

"Fish them out," said Bulkeley.

"On the double, XO. Move us out," ordered Sherertz.

Just then Sherertz noticed that PTs *505* and *508* were closer and moving in quickly.

"Belay that order, XO," said Sherertz.

As he watched men from PTs *505* and *508* pulled the seamen out of the water.

Meanwhile, the *Osprey* had put out the fire. Smoke was still evident, and she was beginning to list to port. The men on deck made short work of jettisoning the deck ammunition in case fire broke out again.

The USS *Chickadee* in the second column, abreast of the *Raven*, moved over next to the *Osprey*, in case the *Osprey* needed firefighting or other assistance. No further fires developed, but by 1816 it was clear that the *Osprey* was going to sink. All remaining personnel abandoned ship to the adjacent *Chickadee*. Several of the *Osprey* men were wounded, and twenty-two were burned. Five enlisted men and one officer were casualties, the first of the D-Day invasion.

As the *Osprey* sank, Sherertz and Bulkeley noted that the destroyer USS *Laffey* left its position a short distance behind them and moved away. They knew it had come in case the *Osprey* didn't sink, so that it could assist with sinking her to keep her from falling into German hands.

~~~

Shortly after midnight, Sherertz and Bulkeley heard the sound of C-47s flying back to England after dropping paratroopers over France. This continued for two hours as each group of thirty-six planes in three-V formation flew by at six-minute intervals. No gun response occurred from the Allied ships, as all ship COs had been briefed.

At 0130, they heard the sound of more planes overhead, this time headed toward France. Because of the low cloud cover and darkness, they couldn't see anything.

~~~

Figure 75. Minesweeper detonating mines while escorted by PT boats on D-Day at Utah Beach. National Archives, history.navy.mil, Photo # 80-G-231649.

The minesweepers finally finished clearing the channels for the troop transports and the rest of the invasion fleet, systematically exploding all those they found. They headed back to Weymouth, leaving the RON 34 PTs behind.

Figure 76. D-Day, RON 34 PT boat close enough to Utah Beach that a wave is breaking in front of it. Jack Sherertz collection.

RON 34 took up positions on the Mason Line to prevent Cherbourg-based E-boats or submarines from infiltrating the western flank of the D-Day invasion. The Mason Line was an imaginary line extending from the French coast 6.5 miles northeast (60 degrees on the compass), to intersect a point 49 degrees, 32 minutes, 30 seconds North and 01 degrees, 06 minutes, 55 seconds West. The PT boats were spaced every five hundred yards, initially located a few miles north of the Saint-Marcouf Islands. Lieutenant Sherertz, Lieutenant Harris, and Commander Bulkeley were aboard the command PT boat, *PT-501*, less than 600 yards from shore.

Finally, Sherertz and Bulkeley began to see the arrival of Force U boats. First the LSTs, pulling their barrage balloons, positioned themselves a few miles off Utah Beach. The USS *Bayfield* was among them. It would be the flagship for the Utah Beach invasion.

Figure 77. USS *Bayfield*, Flagship for Utah Beach invasion. USS *Nevada* visible in the distance on the right. US Navy NationalNaval Archives, history.navy.mil, Photo # 80-G-252377.

Then eight destroyers moved in and anchored. The USS *Fitch* and USS *Corry* were located south of the Mason Line and closer to shore than the Saint-Marcouf Islands. Finally, the big ships arrived (cruisers USS *Tuscaloosa* and USS *Black Prince* closest to the PT boats and cruiser USS *Quincy* and the battleship USS *Nevada* farther to the east).

Figure 78. USS *LST-382*, D-Day. Jack Sherertz collection.

Between 1:00 and 4:00 a.m., the LSTs began discharging their Higgins landing craft, which moved into a holding position just beachward of the LSTs.

"Skipper, doesn't it look like we're too far from the action?" whispered Sherertz to Bulkeley, as the moon shown through a break in the clouds.

They had been ordered to keep the LSTs in close visual range, so that the Germans couldn't slip into a gap between them. At that moment the LSTs were barely visible with binoculars.

"Right you are," said Bulkeley softly. "I'll bet those force 5-6 southeasterly winds blew them off course. Move us closer, Sherertz." Unlike the other boats the PTs had been able to use the Marcouf Islands to help maintain position, so they had not drifted. As the 501 Command PT boat moved slowly southeast, parallel to shore, to close the gap with the LSTs, the nearest boat seaward would move to maintain the 500 yard gap until all the PTs had repositioned.

~~~

At 0520, daybreak, Sherertz could hear the sound of planes coming their way. Then he could see bomb flashes on shore, followed by the deafening roar of detonation, as the Ninth Air Force B-26 Marauders dropped thousands of 250-pound bombs from 5000 feet.

It begins, thought Sherertz to himself.

The resulting bomb clouds obscured Utah Beach, and

the noise was deafening. Then waves of rocket barrages drenched the beach aimed at the German shore batteries.

At 0530, a shell from shore whistled over the PTs on its way to a near miss of the USS *Quincy*, followed immediately by other shore batteries targeting the warships. At 0550, the battleships, cruisers, and destroyers began their broadside shore bombardment. The muzzle flashes were visible all over the bay, and the noise was unimaginably loud. Under cover of this barrage, the LSTs began moving toward shore. Some of them cut loose their barrage balloons, as the cables flapping in the wind were endangering their crews. With the shell explosions lighting up the whole horizon, the sunrise to the east at 0600 went almost unnoticed.

Figure 79. USS *Arkansas* bombarding Omaha Beach on D-Day. US Navy National Archives, history.navy.mil, Photo # 80-G-231250.Wikipedia.

At 0610, planes appeared to lay a smokescreen over the destroyers. The plane that was supposed to obscure the USS *Corry* was shot down, resulting in the USS *Corry* being the only destroyer visible. Directed by spotters on the beach the German shore batteries at Crisbecq, two miles inland, concentrated their 210 mm fire on it. The Corry took evasive action and moved outside of the mine-swept lanes. In short order, it took two direct hits from mines and cracked wide open amidship. The detonations had such power that men were blown fifty feet overboard.

As Sherertz watched with binoculars, he could see one man's lifeless body hanging from the mast.

Figure 80. USS *Corry* hit amidships at 0630 D-Day, image shot from *PT-507*. US Navy National Archives, history.navy.mil.USS Corry sinking. Collection of George K.S. Hardy, crewmember of USS Fitch (DD-462).

"Abandon ship," blared from the Corry PA system. The men jumped overboard and attempted to swim toward the Saint-Marcouf Islands, but the current kept bringing them back to the *Corry*. The ship slowly settled to the bottom, with its mast still above water. One of the remaining crew pulled the body from the mast and replaced it with the US flag.

"Sherertz, if no one comes to the *Corry's* aid in the next few minutes, we'll help out until they get here," said Bulkeley.

"Aye, aye, sir," said Sherertz.

Just then they saw *PT-199* roaring toward it from the direction of Utah Beach. Meanwhile, the USS *Fitch*, USS *Hobson*, and USS *Butler* dispatched their whaler launches to her aid as well. Over the next two hours, these four ships pulled out 208 survivors from a crew of 264, in spite of continuous shelling from the shore. *PT-199* rescued 61 men and transported them to the USS Fitch, then headed back to its station at the USS *Bayfield*.

Meanwhile, twenty landing craft vehicles (LCVPs), each carrying a thirty-man team, were underway toward Utah Beach. When they reached a point several hundred yards from the beach, smoke projectors were set off, visible for miles. Immediately all the destroyers and battleships behind them stopped firing, and deathly quiet ensued. The gusting southeasterly wind slowly cleared some of the smoke and sand from the air, and the beach became visible. Closer to shore the LCVPs encountered steel and concrete pikes, steel tetrahedrals like giant jacks, and hedgehogs. Unable to move further forward, the LCVPs discharged their men at almost exactly 0630. Immediately the quiet was broken by the sound of machine-gun and rifle fire from Utah Beach. Amid the swirling smoke and the rain of bullets, the men trudged the last hundred yards to the beach.

Bulkeley and Sherertz's PT boat remained a few hundred yards off shore, with its bow pointing toward England to handle the occasional wave that broke over them. In this position they could make sure that no German frogmen or small boats came between them and shore. Each of the four gunner's mates was outfitted with 7x50 binoculars, looking around 360 degrees to be sure they weren't missing anything. Sherertz and Bulkeley had their own binoculars and were watching the shore batteries, in case evasive maneuvers became necessary. Also on their boat was one of Commander Ford's photographers, busily snapping photographs.

Sherertz could easily see the US soldiers moving toward shore through his binoculars. Initially it looked almost like a drill because the German guns on Utah beach were small in number. But the guns were accurate and soldiers began falling, some silently, many screaming, red circles of blood forming around them where they landed. He could see one eviscerated soldier lying on his back and another, who had an arm shot off, holding onto a concrete pike with his good arm while he screamed for help. Suddenly Sherertz's mind was filled with images from the Pearl Harbor attack.

After what seemed like an eternity to Sherertz, but was actually just a few minutes, the Utah invasion forces had taken out the few German machine guns and begun moving inland. No casualties had occurred on board the PT boats.

The lack of initial resistance allowed the second wave (amphibious tanks), third wave (dozer tanks), and fourth wave (engineer combat battalions) to land with minimal interference. They lost only three LCTs and one LCM. As a result, they were able to clear the beach of all obstructions by 0945 and immediately began working on creating gaps in the seawall for heavy equipment exit points.

Soon thereafter, shelling from inland German forts began to hit the beach. Fortunately, most of the men and most of the vehicles had already moved off the beach. All day long, the LSTs discharged men in an endless stream.

As the daylight faded, Bulkeley took one last look around with his binoculars. Seemingly satisfied that all was in order, he turned to face Sherertz.

"I'm going to head up on the bow and see if I can see the full moon through the clouds and try and catch a few winks," said Bulkeley. "You set the watch, Sherertz."

Everything was dead quiet until near midnight when a plane flew over, that the Allies had not been briefed to expect. In short order, night turned to day as every ship with an anti-aircraft gun opened fire. The plane escaped, but the excitement kept most everyone awake the rest of the night.

Figure 81. Prayer Services to be held, 6/6/44. Roanoke World-News clipping. Can be found at www.AHeroAmongMillions.com.

Figure 82. The president's prayer, 6/6/44. Roanoke World-News clipping. Can be found at www.AHeroAmongMillions.com.

Figure 83. D-Day Rube Goldberg cartoon. Roanoke World-News clipping. Can be found at www.AHeroAmongMillions.com.

USS Glennon, USS Rich

Figure 84. USS DE Rich sinking after being hit by mines. US Navy National Archives, history.navy.mil, Photo # NH 44312, Wikipedia.

June 8, 1944

On D-Day plus two, the Third Battalion Twenty-Second Infantry soldiers were trying to work their way north from Utah beach toward Cherbourg. Their mission was the methodical destruction of all beach defenses between Utah Beach and Cherbourg. However, it had been very slow going, and they had only advanced four thousand yards in two days. Now they were unable to move because of shelling from the 210 mm guns at Crisbecq. They radioed for naval gun support.

"Pinned down. Can't move. Some of you tin cans give us some cover fire."

"Skipper, we're right here," yelled Crist over the radio. "Give us a shot at the Jerries."

"It can't hurt," said Bulkeley, standing next to Sherertz.

The closest naval guns to their shore position were those of RON 34, still anchored along the Mason Line.

Bulkeley called the *Bayfield* and requested permission to fire. "All right men, we've been cleared for action," he said. "Move in close to our position and give 'em hell. Sherertz will give you the vector for Crisbecq."

"Two hundred thirty-five degrees southwest," said Sherertz. "Range 2.5 miles. Fire when ready with your forty millimeter."

The 40 mm cannons had a range of four miles and delivered their shells with accuracy, but they had no effect on the cement bunkers housing the 210 mm guns.

"This is Little Boy calling Big Brother," said the radio operator on the *PT-504* boat that Bulkeley and Sherertz were using that day as their command boat. "We need bigger guns. Is anyone available?"

"Little Boy, this is Big Brother. Help is on the way."

"They're coming, Skipper," yelled the radioman from below deck.

At 0800, the PTs saw the destroyer USS *Glennon* moving toward a position near them to open fire with its five-inch guns. The PTs watched closely, waiting for the coming salvo.

Boom, echoed a loud explosion from underneath the *Glennon*.

The port quarter of its stern lifted high in the air, with a rift visible in the deck as it settled back down. The stern section took on water and sank to the bottom. Within minutes, men from the damaged compartments were pouring out into the water, most badly wounded and yelling for help. The *Glennon* put her whaleboat into the water to pick up survivors.

At 0806, two sweeps and the US DE Rich were dispatched to the *Glennon's* aid. All three went to general quarters and proceeded at speed to the site of the floundering destroyer. Upon closing, the sweeps moved into position to try and tow the *Glennon*, and the Rich dispatched its whaleboat to offer assistance.

"Do you need assistance?" yelled the Rich commanding officer, Lt. Comdr. Edward A. Michel Jr., via bullhorn.

"Negative," replied Commander Clifford A. Johnson, skipper of the *Glennon* through his bullhorn. "Clear area cautiously. Live mines."

The *Rich* whaleboat turned around and headed back toward the *Rich*, picking up survivors in the water as it went. Meanwhile, the *Rich* slowly moved around the *Glennon's* stern toward the starboard side, preparing to depart.

Boom. A projectile from a shore battery exploded just ahead of the *Rich*, sending a geyser of water over a hundred feet in the air. No apparent damage was done.

The *Rich's* whaleboat tied up at the *Rich* portside boat fall, preparing to board.

Boom. A mine exploded directly under the *Rich's* stern, lifting it high in the air and blowing off seventy-five feet of the fantail. Men and deck supplies were blown fifty feet into the water, followed by loud screams. The explosion and resulting wave rocked the whaleboat, nearly capsizing it. The men cut loose from the boat fall, fearful of being swamped, and headed back into the smoke coming from the *Glennon*.

Lieutenant Whorton, skipper of *PT-508*, was watching the *Rich* when the explosion occurred and radioed Bulkeley for permission to provide aid. With permission granted, *508* roared toward the *Rich*.

"Sherertz, mobilize the *502* and the *506*, and we'll all go," ordered Bulkeley. The other three boats roared after the *508*, coming to a banked stop adjacent to the *Rich*.

"Do you need help?" yelled Bulkeley through his megaphone.

"No help needed," replied the *Rich*.

Boom. Another mine exploded under the forward hull, reducing the forward section to complete wreckage. The blast showered the PTs, now only fifty yards away, with shrapnel, water, and body parts. Now there were more seamen in the water, and the ship was sinking faster.

"Men overboard. All hands on deck," yelled Bulkeley.

The *Rich* whaleboat docked at the boat fall, and the men scaled the gangway to provide assistance on board.

Boom. A third mine exploded amidships, and the *Rich* began sinking fast.

"All PTs tie up and bring survivors aboard," yelled Bulkeley over the radio.

On board the *Rich*, they found dead bodies and body parts, as well as survivors with all conceivable injuries, many pinned under debris. In the few minutes they had, they freed all survivors possible and transported them to the PTs. Those trapped by debris, but still alive, were given large doses of morphine to ease their impending drowning. They found Lieutenant Commander Michel and his XO on the bridge with their legs broken, wielding pistols and threatening to shoot anyone who tried to take them off the ship. Bulkeley knew Michel from service together on the USS *Indianapolis* and managed to talk him out of shooting anyone, so they were able to rescue both men. By the time they carried them onto the PT's deck, the *Rich's* decks were awash.

"Abandon ship," ordered Bulkeley, and there was a scramble to board the PTs as the *Rich* sank. Suddenly, two men popped to the surface as the *Rich* disappeared below the water. R. W. Gretter of the *504* boat and P. E Cayer of the *506* had been so busy trying to save more men that they didn't hear the abandon ship order and went down with the Rich, albeit briefly. They both sported huge grins as ropes were thrown their way.

"Look there, one more man overboard," yelled a seaman on the 508 boat.

"Throw him a line," yelled another.

"Never mind, fellows. I have no arms to catch it," said the seaman, and he disappeared beneath the waves.

"Sparks, I need a status report from the other boat captains about readiness to move out," yelled Bulkeley.

"All boats ready and able," yelled sparks a few minutes later.

Sherertz and Bulkeley looked around. They had rescued sixty-nine men, most wounded, some dying, and almost all of them still on the PT boat decks.

"Well Sherertz, our men have seen death up close for the first time," said Bulkeley.

"It won't be their last," said Sherertz.

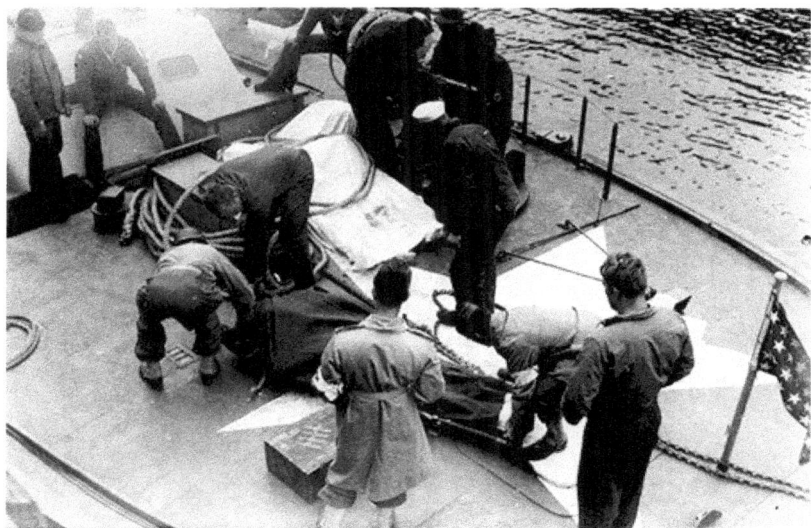
Figure 85. *PT-503* with casualties from USS *Rich*. Jack Sherertz collection.

PT Boat Guests

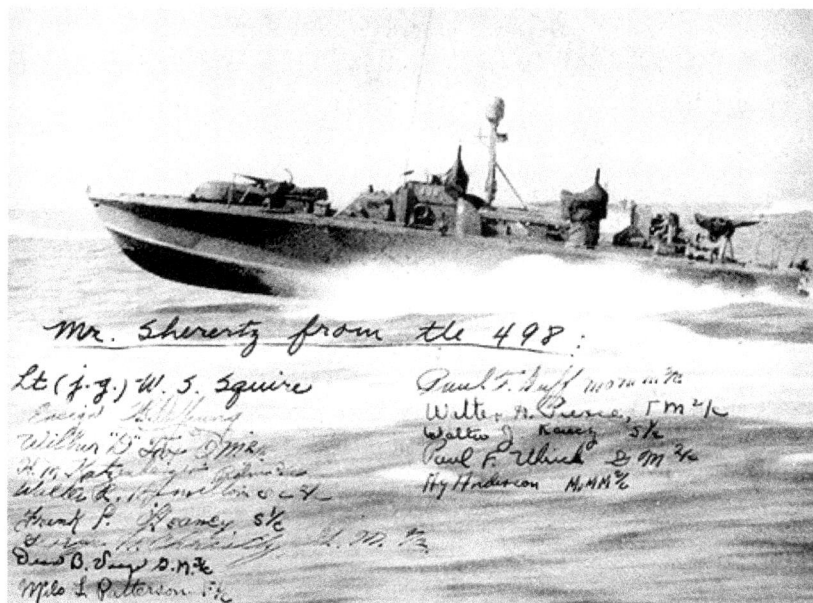

Figure 86. PT- 498. Jack Sherertz collection.

June 10, 1944

On D-Day plus four, *PT-498*, the Jolly Roger, was patrolling the Channel just northeast of Cherbourg in midmorning. The rest of RON 34 was similarly deployed in other parts of the Channel. Lieutenant Sherertz stood on the bridge with boat CO Lieutenant Junior Grade Squire and boat XO Ensign DeYoung. Clouds were visible over France, and they had been cruising easily at twenty knots for half an hour without any sightings. Then from a distance they heard the unmistakable sound of an airplane engine. Everyone on deck started searching the sky.

"Plane ho at one o'clock!" yelled the gunner's mate. "I think it's a Kraut."

Sherertz and Squire pulled up their binoculars and took a closer look.

Figure 87. Arado Ar 196. Wikipedia.

"I'll be damned," said Sherertz. "It's a Kraut surveillance plane, an Arado Ar 196. They used to be attached to heavy cruisers and launched from on deck. But ever since the Brits wiped out their cruisers, they've been flying from land bases. This one probably came from either Cherbourg or one of the ports on the western side of the Cotentin Peninsula. They can't see any troop movements right now because of the clouds, so they've come to see what kind of boats they can find. Look alive, everyone. They've got machine guns, twenty-millimeter cannons and bombs. Be ready to fire or use evasive tactics!"

For the first few minutes, the plane flew around slowly in big, lazy circles a few thousand feet above them. Then suddenly they heard the sound of a plane coming from the direction of England. The Kraut plane evidently heard it, too, because it turned and disappeared into the clouds, an English fighter plane in hot pursuit. Soon thereafter, they heard machine-gun fire and the sound of a plane going into a dive. The Kraut plane broke out of the clouds, streaming smoke behind. It tried to gain altitude but then began a slow roll to the left that was headed toward a crash landing. Two men opened their cockpits and leaped out, followed by parachute openings. Just above them, the English plane burst out of the clouds in time to see the Kraut plane crash into the water, and then took off looking for more prey.

"Let's go pick up those Germans," said Sherertz to Squire and DeYoung.

DeYoung roared off after the parachuting pilots. When they arrived, only one man was visible on the surface, still trying to get out from under his parachute. All men on the PT were on deck with rifles or pistols pointed at the one pilot in the water. After he got out from under the chute, he raised his arms slowly, showing he was unarmed, and said something in German.

One of the men shot into the water near him.

"I surrender," said the pilot in perfect English, to the surprise of everyone on deck.

The men edged closer with their guns and looked quite menacing. Suddenly Sherertz pushed through the men and turned to face them.

"I want him alive! Is that clear?"

"Aye, aye, sir," came the loud response.

"Where's the other man?" Sherertz yelled.

"I think he was shot and sank soon after we hit the water," the German yelled back.

"Get him on board and down below," said Sherertz to Squire.

In short order, the German was pulled on board, searched for weapons, and taken below.

"Why don't you see what you can learn from him?" said Sherertz to Squire. "DeYoung and I will get us back to base."

Squire smiled at Sherertz. "My pleasure." A short while later, Squire returned to the bridge. "We won't be getting anything out of him about their military except name, rank, and serial number. That briefing you and I went to suggested that there was an Arado Ar 196 base somewhere southwest of here in the Bay of Biscay, but he won't own up to it. However, I did learn something else interesting. He's a banker. He's down there giving out financial advice and even promises to help one of the men with some back taxes. I think he actually knows what he's talking about."

"I'm not surprised," said Sherertz. "We'll send him back across the Channel and see if Ike's men can get anything out of him."

"And one more thing," said Squire. "He wants you to have this." He held out his hand and opened it to reveal a Luftwaffe patch with an eagle holding a swastika in its talons. "He pulled it off his uniform and said to give it to you for saving his life."

Figure 88. Kriegsmarine EM tunic eagle coastal artillery patch.
Jack Sherertz collection.

"Let's get back to base so we can get him off our hands," said Sherertz as he accepted the patch.

~~~

"Lieutenant Squire, I thought you might be interested in the message I just got from Ike's interrogation unit," said Sherertz. "It turns out that our so-called pilot was not a pilot. The pilot was killed. Our prisoner was only claiming to be an officer, hoping to get better treatment. I think he did."

"How did they know?" asked Squire.

"First off, he didn't know shit about the airplane," said Sherertz. "But mainly it was that patch he gave you. It was a navy enlisted man's patch, coastal artillery, not a Luftwaffe pilot patch. This is a pilot patch," he said, holding up a picture.

~~~

Figure 89. King George VI arrival for PT transport.
Jack Sherertz collection.

June 12, 1944

King George VI, Winston Churchill, British Admiral Ramsey, and US admirals Strubel and Kirk arrived at the Plymouth dock for an inspection tour of the invasion beach areas.

Lieutenant Sherertz greeted them. "Gentlemen, Lieutenant Aldridge will take you anywhere you need to go. Let me know if I can be of further assistance."

"Aye, aye, Lieutenant Sherertz," said Aldridge, as Sherertz turned to leave.

Aldridge, Lieutenant Sherwood, and Ensign Sharkey behind him shot Sherertz a look, as if to say, "We can't believe you're sticking us with all of the high brass."

Figure 90. *PT-504* transporting King George VI, Winston Churchill, and four admirals to inspect Normandy Beach after D-Day. Jack Sherertz collection.

~~~

Figure 91. World War II Bosun's chair. US Navy
National Archives, history.navy.mil, Photo # 80-G-320987.

Sherertz watched from the bridge as Commander John
Ford was transported in a bosun's chair dangling from a
rope spanning the gap between the USS *Augusta* and
Commander Bulkeley's command PT. Because of the fifteen
- to twenty-foot swells, the bosun's chair went up and

down like a yo-yo, alternately dunking Ford or throwing him high in the air. The dunkings were making it very slow going and making Ford look increasingly like a drowned rat.

When Ford finally arrived over the PT, Sherertz thought he looked awful. He hadn't shaved or changed his clothes in days, and the water made everything look worse. The seamen struggled to get him down, but the rope used to tie him into the chair was knotted tight, and they began looking for a knife.

Meanwhile, Sherertz came down and used the marlinspike that he had saved from Midshipmen School to undo the knot in a matter of seconds. "I live to serve," he said to Commander Ford, with a great flourish.

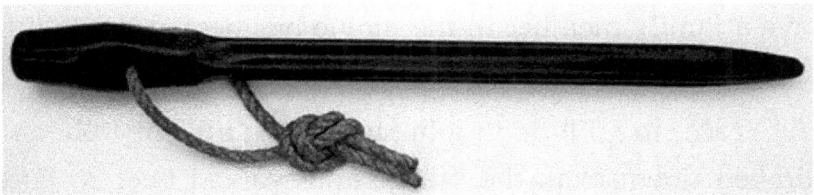

Figure 92. Fit or marlinspike from HJS USS Arkansas training voyage.
Jack Sherertz collection.

Commander Ford looked like he was about to bless him out. Then he started laughing. "Let's go see the Man," he said to Sherertz.

"The Man is on the bridge with the boat CO," said Sherertz. He dutifully followed Ford up to the bridge.

"Commander Ford, welcome aboard," said Bulkeley.

"It's *John* or *Jack,* not *Commander Ford!*"

"Jack, meet me in the captain's stateroom in thirty minutes, and I'll brief you on our PT boat operations," said Bulkeley. "Meanwhile, Sherertz will be happy to answer any of your questions."

Sherertz leaned in and said, "John, I can't believe you brought me here just to babysit this crazy man."

"Jack, it goes with the territory. If you're going to lead, sometimes you have to entertain the brass, which I understand you dodged nicely yesterday."

"Guilty as charged," said Sherertz. "But I have more important things to do then hang around Commander Ford."

"Who knows?" said Bulkeley. "Perhaps some day you'll have a family member in the movie business and you'll be able to say, 'I knew the great movie director John Ford.'"

"Yeah, sure. I'll do it, John, but I won't like it." Sherertz climbed down from the bridge and walked over to Ford. "Welcome to our command PT, Jack. What can I help you with?"

"Good to see you, Jack," said Ford. "Why don't you give me a tour and tell me how everything works around here."

Sherertz rolled his eyes and started walking.

"Jack, I know you don't like me very much, but I think if you'll give me a chance, you'll find that I'm not such a bad guy," said Ford.

"We'll see," said Sherertz.

At dusk, they spotted an E-boat near Cherbourg Harbor. Bulkeley ordered the boat to engage, and soon both boats were doing nearly forty knots, running parallel to each other and firing all guns. Sherertz and Ford were at the stern, looking at the depth charges when the sighting first occurred. Before Sherertz could say anything, Ford climbed up to the bridge to watch Bulkeley. In spite of the earsplitting sound of the twin .50s and bullets flying all around them, Ford didn't flinch. After the firing was over, Ford climbed down and started talking with the gunner's mates, asking them questions about difficulties with shooting at speed.

Bulkeley and Sherertz watched him from the bridge. "He's a lot tougher than he looks," said Sherertz.

"He's a lot smarter than I gave him credit for," said Bulkeley. "He really understands exactly what we're up to and what it's going to take to win this war. Jack, you think you can cut him a little slack now?"

"Aye, aye, sir," said Sherertz with a smile.

Storm

June 19, 1944

Lieutenant Sherertz was using *PT-500* as his command PT while patrolling the Channel Islands, when he received a radio communication from Commander Bulkeley.

"Sherertz, the worst storm in half a century is bearing down on you. You have only a few hours to find a safe tie-up. Break formation now and get to one of our safe harbors. That is all."

Sherertz knew that Bulkeley wouldn't give that kind of an order unless he was sure. With any ordinary storm, he would have let them ride it out and continue patrolling. That gave him grave concern because their present location was just south of Jersey Island. That meant if they swung wide enough around Jersey and Guernsey Islands to avoid the Jerry guns and similarly swung around the guns at the mouth of Cherbourg Harbor, they would have to traverse over a hundred miles of open ocean to get to the Omaha Beach mulberry.

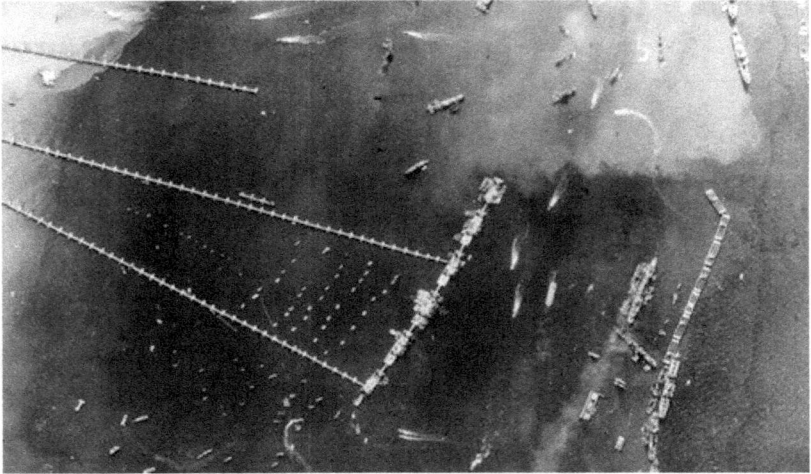

Figure 93. Mulberry artificial harbor off Arromanches in Normandy, September 1944. Blockships are seen at the end of the long docks oriented parallel to shore to act as a breakwater. Wikipedia.

After the D-Day invasion, US and English naval engineers had constructed static breakwaters off each of the invasion beaches called "corncobs," made of scuttled block ships. The calm water beachside of each corncob was referred to as a gooseberry. Two of the gooseberries, Omaha Beach and Gold Beach, were upgraded to full-fledged harbors or "mulberries" by adding floating outer steel breakwaters called bombardons, reinforced concrete caissons called "Phoenixes," floating piers or roadways named "whales," and pierheads named "spuds."

"All you flying deuces out there, this is the skipper. Get to our berry breakwater, flank speed. A major blow is coming, and we need to be tied up when it hits." Sherertz looked around. Paradoxically, the water was almost at a dead calm. *That will help*, he thought to himself.

"All right, Ensign Allen, flank speed," said Sherertz to the XO of *PT-500*.

"Flank speed, aye, aye, sir," replied Allen as he reached for the throttle of each of three engines and moved them all the way to full. In three quick jumps, they hit full speed, planing well out of the water. The sound of three 1500 horsepower engines, fully unleashed, was quite impressive to hear, not only on the PTs, but to the enemy as well. As they swung by Jersey and Guernsey Island, the 88 mm guns fired a few shells in their direction, hoping to get lucky. But the planned course was well out of range, so the shells all fell short.

By the time they passed Cherbourg Harbor, the wind started to pick up. Ten knots, twenty knots. Then the rain began. By the time they got to the Omaha gooseberry, the wind was gusting to fifty knots with horizontal rain.

"Call the Omaha mulberry harbor control and check on its status," said Sherertz.

"They're full up," said XO Allen.

"I don't think we have time to get to the Gold mulberry," said Sherertz. "We're going to anchor beachside of the Utah blockships. Tie 'em up, fore and aft, and make it secure. Rig the docking cushions over the side to protect the hulls."

This was all accomplished in a few minutes, in spite of the wind and rain. Sherertz heard back that each boat was secure and then spoke one last time over the radio. "Everyone get below, so no one gets washed overboard. Batten down all hatches and anything else that's not fastened down."

It was a good thing they had not tried to reach the Gold mulberry, because minutes later the first of a series of thirty -foot waves came at the beach and broke over the block ships before crashing down on them. The first large wave that broke over *PT-503* snapped its mooring line. Fortunately, a nearby tug was there to provide them with a steel cable before they drifted away.

Sherertz was in the radio room below deck when XO Allen walked in drenched. "What happened?" asked Sherertz.

"The wind blew one of the hatches open and soaked Lieutenant Kennedy and me before we got it tied down again," said Allen.

"Stan, let's do a quick walk-through and see if anything else needs to be tied down more securely," said Sherertz.

In spite of dungarees being stuffed in ventilator shafts, water was pouring out of each of them below. When Sherertz and Allen walked into the galley, pots, pans, utensils, cans of food, and the cook were sloshing back and forth with each wave. Sherertz and Allen helped try to put things back into compartments and then moved on to the engine room. As they staggered into the room, a battery broke loose from its shelf attachment and nearly hit Allen in the head. As soon as it hit the salt water, it started smoking. As the storm progressed, they were thrown back and forth, crashing into the walls. Everyone had bruises from repeated falls. One of the motor mechs had wrapped himself around a ladder and was praying loudly. Sleep was impossible.

The rolling was so bad on *PT-509* that steel helmets ended up on the roof of the dayroom and then fell into the water without hitting the deck.

~~~

Finally, after three days, the worst of the storm was over, and Sherertz called Bulkeley to get new orders. "Skipper, where do we head?" he asked.

"Come back to base to get checked for damages. You're lucky you didn't get to the Omaha mulberry. I'm told a good part of it washed away during the storm, and a number of the ships in the harbor sank."

Drawing Fire from Cherbourg

Figure 94. PT boats in Channel waters. Jack Sherertz collection.

June 25, 1944

Jack Sherertz stood on the bridge of *PT-510* with Bulkeley as they motored toward Cherbourg Harbor. "What's the story, Skipper?" Sherertz asked.

"The Allied land forces are moving on Cherbourg and planning on capturing it as soon as possible," said Bulkeley. "There is one major obstacle: the eighty-eight millimeter guns at the fort in Cherbourg. Our orders are to draw the fire of those guns, so our battleships can see them and wipe them out. Sound like fun?" Bulkeley grinned at Sherertz.

"You really are crazy," said Sherertz.

"All right, XO, head toward the fort," said Bulkeley.

PTs *510* and *521* charged to within a thousand yards, to no avail.

"Closer, and be ready to move fast," said Bulkeley.

Both PTs sped to within two hundred yards of the fort, and still got no response.

"Fire a burst from the fifties," said Bulkeley.

Tat, tat, tat, tat, tat, tat, tat, tat, tat, went the .50s.

Boom, boom, boom went the 88s in response, immediately giving away their position. Suddenly a near miss of *PT- 521's* stern caused all three engines to stall.

"Crank up the smoke, and *circle 521,*" commanded Bulkeley.

PT-510 circled *PT-521,* with smoke trailing, but it became increasingly difficult to see *521* and not collide with it. Finally, after a few minutes of frantic work by the motor mechs, the engines roared back into action.

"Let's get the hell out of here," said Bulkeley. "Would you agree, Lieutenant Sherertz?" He smiled.

"Most definitely," said Sherertz. "Now, if not sooner. And by the way, why do you keep inviting me to such fun occasions?"

"Let's just say, I want to be sure you're ready for the next step," said Bulkeley. "CQ, PTs are departing Cherbourg."

Within seconds of Bulkeley's radio message, they heard the sound of battleships firing their big guns. They turned to look behind them at Cherbourg and saw the rain of shells begin. When they turned back around, toward the channel, the battleships came into view. Jack instantly recognized among them the profile of the USS *Nevada*. As he watched, the *Nevada* was straddled multiple times portside and starboardside by 88 mm shells. Continuous weaving by the crew kept her from being hit. *Give'em hell, girl*, he thought, with much emotion. In the minutes it took them to leave the harbor, the fort was largely destroyed, the 88 mm guns silenced forever.

The next day within Cherbourg Fort du Roule was captured, the dominant location of German defenses. The harbor fortifications and the arsenal surrendered June 29th. Immediately thereafter the allies set about fortifying the city, in case the Germans counterattacked, and began clearing the harbor of mines and debris so that Cherbourg could be used as an Allied port.

Figure 95. Fort at Cherbourg after Allied shelling. Jack Sherertz collection.

Sherertz, CO of RON 34

Figure 96. Lt. Sherertz on board PT. Jack Sherertz collection.

July 16, 1944

Most of RON 34 was patrolling the Channel Islands when Lieutenant Harris arrived at the dock at Portland Bill. He and Lieutenant Sherertz did a quick inspection of the remaining PTs. Then they shook hands and just that quickly, Harris, in his new position as Commander of all four Channel PT squadrons, officially turned over the command of RON 34 to Sherertz.

"Lieutenant Sherertz, now that we have captured Cherbourg and cleared the harbor, I want you to get over there and pick out a site for the headquarters for our new naval base, which by the way, you'll be commanding."

"Aye aye, sir," said Sherertz as he saluted. *That didn't take long*, said Sherertz to himself.

Bulkeley was now captain of the destroyer USS *Endicott*. He had left without interacting with Sherertz. The day after Sherertz assumed command of RON 34, a messenger brought Lieutenant Sherertz a handwritten note.

> Sherertz, it's been great working with you. I want to encourage you to stay in the Navy. I think it would be a great career for you.
>
> Best wishes.
>
> John Bulkeley.

~~~

Figure 97. Amphibious vehicle coming aground in Cherbourg. Minesweepers clearing mines in the background. Jack Sherertz collection.

August 4, 1944

After the Germans surrendered the Allies immediately began moving troops and supplies into Cherbourg to establish bases to protect the harbor as a supply route for the assault on Germany. RON 34 CO Lieutenant Sherertz was now the Commander of the newly established Cherbourg Naval Base. Sherertz, Squadron XO Lt. J. J. Daniel, and a motor mech had come ashore via a duck amphibious assault vehicle to look for a site for a PT boat base. They pulled up to a road and flagged down an MP in a jeep.

"Which way to the docks?" asked Lieutenant Sherertz.

"What'cha lookin for, Lieutenant?"

"We're scouting a base location for our PT squadron," said Sherertz. "We're looking to anchor twelve boats and set up a dry dock for repairs. Any suggestions?"

"I think I know an area that might suit your needs," said the MP. "Hop in and I'll give you a ride. Where you from, Lieutenant?"

"Roanoke, Virginia," said Sherertz. "Where you from?"

"I'm from the West Coast. Portland. Ever been there?"

"Never," said Sherertz. "What's it like?"

"Wet!" said the MP. "Kind of depressing, actually. But a lot of people like it, because it's so green."

Figure 98. Cherbourg ruins. Jack Sherertz collection.

They didn't go a long distance, but they had to take a circuitous route because of all of the damage caused by Allied and German shelling and German demolition of buildings. The MP dropped the three off on a corner right next to the harbor.

Figure 99. Cherbourg harbor. Jack Sherertz collection.

"Thanks for the ride," said Sherertz.

"I see some areas that may work for a dry dock," said the motor mech. "I'll check them out."

Lieutenant Sherertz and XO Daniel headed toward the dock.

"Jack, I've heard from just about everyone else about what they want to do after the war," said XO Daniel. "How about you?"

As they walked across the street, they cut in front of a line of Army trucks. Their easterly direction suggested they were about to leave town.

"Haven't thought about it much," said Sherertz. "I've been too busy keeping this bunch of misfits out of trouble. However, I have been thinking a little bit about the possibility of staying in the Navy. Bulkeley has been pushing me to re-up after the war."

"Incoming," yelled a voice from behind them, as the high pitched scream of an 88 mm shell was heard coming their way.

"Booom."

Knocked off their feet by the explosion behind them, they turned and looked at the street corner they had just left. It was gone, except for a big hole. A truck they had just passed had been hit pretty bad by shrapnel. The driver hung out of an open door. He looked dead. Some civilians nearby were hurt and bleeding.

"Holy shit, Skipper," said Daniel. "That could have

been us."

Everyone near the dock came running, pointing and talking in an animated fashion. Some of Sherertz's squadron, who had been riding behind them, unloaded to see if they could help. They put the truck driver on a stretcher and took him toward a medic vehicle. Another man from the squadron got into the truck and drove it out of the middle of the road, then caught up again with the convoy.

Figure 100. US Army truck in Cherbourg.
Jack Sherertz collection.

"That was too close," said Daniels.

"J. J., perhaps God has something in mind for us other than war," said Sherertz.

~~~

Daniels came into the makeshift RON 34 barracks

headquarters set up temporarily in a tent.

"Jack, you know how we saw one of our men drive that Army truck out of the middle of the road?" Daniels said.

"Yeah, I remember," said Sherertz.

"Well, it was Torpedoman Easterly from the 503," said Daniels. "And you know what else? He didn't come back. He's AWOL."

~~~

Ensign Schuster walked into the makeshift brig, a room with a guard in a mostly bombed-out building.

"Hey, Easterly, why did you come back?" Schuster asked.

"They figured out I wasn't one of them."

"Why did you leave?"

"We hadn't seen any action since D-Day, and I came over here to fight. I thought maybe the Army would let me fight at the front. It's a good thing the skipper doesn't know about my still."

"What makes you think he doesn't?" said Schuster to a wide-eyed Easterly. Ensign Schuster shook his head and walked out.

Standing just outside was Lieutenant Sherertz, who went in as Schuster left. "Well, Easterly, we got ourselves a situation," said Sherertz.

"What situation?" asked Easterly.

"Lieutenant Harris wants you court-martialed and sent

stateside to a military prison."

Easterly looked devastated.

Sherertz paused to let it sink in. "I, on the other hand, think that you would be better off fighting the war. What if I court-martial you, dock you a month's pay, and keep you in the brig as though waiting to be sent stateside. Then when I need a replacement on one of my boats, I'll tell him that I can't get anyone else and I would like to use you instead."

"I can do that," said Easterly quickly. "Thank you, sir."

"If you speak a word of this to anyone, I'll deny it and you will be sent stateside. Do we have an agreement?"

"Thank you, sir. I'll never forget this."

Figure 101. Sailor finds Navy Dull. Roanoke World-Newsclipping. Image can be found at www.AHeroAmongMillions.com.

PT- 509

Figure 102. USS DE Maloy. Wikipedia.

August 8, 1944, 1700 hours, Cherbourg Harbor

Lieutenant Sherertz arrived at the dock where the USS destroyer escort USS *Maloy* was anchored. The USS *Maloy* was launched 8/18/43. She initially escorted troop carriers, then was involved with amphibious training prior to D-Day and then, after D-Day, it was assigned to assist RON 34 in patrolling the channel. The more powerful radar unit on the *Maloy* would direct RON 34 on night maneuvers. Six PTs would travel with the *Maloy* until they reached their mission destination.

As Sherertz walked toward the *Maloy* gangway, he passed his PTs. He noted with some pride that they were all ready to go and looked shipshape.

"Yo, Skipper," yelled one of the men. "You think there's going to be any action tonight? We're all tired of ambulance duty."

"Be careful what you wish for," replied Sherertz.

Sherertz walked up the gangway of the *Maloy*, saluted the sentry, who recognized him and let him pass, and found his way to the radar room. As expected, he saw British officer Lt. Commander Peter Scott, on loan to the US Navy from the British Royal Navy as vector controller. Scott's father was the well-known Antarctic explorer.

"How's my favorite limey?" asked Sherertz with a British accent.

"Union Jack! Come to chew the fat?" asked Scott as he turned from the radar screen.

"They told me on deck a real Royal Navy officer was in the radar room, but I see they are wrong."

"He just left. If we wait, maybe he'll come back. How 'bout a cuppa joe and a butt?" said Scott with a New York City accent.

"I'd rather have a little grog."

"For you, sir, I have brandy."

"Now you're talking, but keep the bottle out of sight. Our higher-ups frown on drinking on duty, but they'll forgive it coming from a limey."

"We'll limit it to one." Scott poured a brandy for Sherertz and himself and then put the bottle away, out of sight.

"How's Captain Cazalet and the HMS *Saumarez*? asked Sherertz. "That was really fine of him to invite all of our squadron officers on board for a drink. Especially after we nearly blew the D-Day surprise attack."

"Bored. Since the D-Day attack, there hasn't been much action. He sends his regards."

"I know the feeling."

"How're your lads doing?"

"Five by five, but bored. One of my men went AWOL."

"We've had similar problems. Since we sank all their big ships, things have been pretty slow."

"Can you believe the names they gave our boats tonight?" asked Sherertz. "The Tunney Group has *Kipper* – PT-500, *Herring* – PT-503, and *Haddock* – PT-507. The Barracuda Group has *Crab* – PT-506, *Crayfish* – PT-508, and *Lobster* – PT-509. And Sturgeon for the *Maloy*. Who thinks up this stuff?"

"Perhaps we should have some fish and chips," quipped Scott.

The small talk continued as they headed to their mission location for the evening. Their mission tonight was to patrol the sea between Jersey and Guernsey Islands. Both islands were German occupied and heavily fortified. They were to engage only if the Germans made a move.

The seven boats passed to the north and west of

Guernsey. At 1945, they received a hail.

"This is *Crayfish* calling *Sturgeon*. This is *Crayfish* calling *Sturgeon*. Do you copy?"

"We read you, *Crayfish*."

"Number two engine broke a propeller shaft. We're heading back to base."

"Roger that. Proceed to base. *Sturgeon* out."

At 2210 they reached their surveillance positions. Tunney Group was located just southwest of Guernsey, Barracuda Group just northwest of Jersey, and *Sturgeon* patrolled a six mile north-south line between Jersey and Guernsey Islands, turning on the north end on the hour and the south end on the half hour.

~~~

Sherertz walked over to the radar screen and pointed to several blips. "Peter, are those three blips up near Guernsey Island the Tunney Group, and the two blips down near Jersey Island the Barracuda Group?"

"Righto, Jacko."

A seaman ran into the room and handed Scott a message.

"Blimey! Thank you, Chief Baker. Jack, I think tonight is about to heat up. We just received Ultra information that a Jerry coastal battery is about to relocate from Saint Peter Port on Guernsey to Saint Helier on Jersey. Two small freighters with eight escort vessels. Fog coming fast.

Departure imminent."

~~~

For most of the night, nothing happened. Then early the next morning, at 0450, eight blips appeared on the radar screen coming from Guernsey Island.

"Here we go," said Scott. "Eight bogeys heading your way. Course 075, speed fourteen knots. Tunney, do you copy?"

"Aye, aye, sir," said Radioman Fisher.

"What's your visibility?"

"Zero," said Fisher. "We're fogged in."

"I'll get you close, and you can fire your torpedoes. Your vector is now 120. Get ready," said Scott.

"Our radar's picked 'em up," said Fisher.

At 0535, the radar screen showed three PT boats rapidly closing on the eight bogeys. At two thousand to three thousand yards, they were now in range.

"Hold course. . . Fire!" said Scott.

Boats *503* and *507* fired their starboard torpedoes, and *500* mistakenly fired both. The Mark 13 torpedo had a top speed of thirty-eight miles per hour. Two minutes should have been enough for torpedo impact.

"Any contact?" asked Scott.

"Nothing," said Fisher.

"Tunney Group, return to position one," said Sherertz. "Barracuda Group, prepare for vectoring toward bogeys heading in your direction."

"Barracuda Group, your vector is 340 degrees northwest," said Scott. "You should close on target in less than five minutes. I'll let you know when you're in range."

"Visibility one hundred yards," said Lieutenant Crist. "Visibility twenty-five yards."

"Prepare to launch," said Scott.

"Ready, fire," said Crist. "Fish away . . . No explosions. Let's circle around for a second run."

"That's it," said Scott. "Five degrees more. All right, they're dead ahead. Fire when ready."

"Two fish away . . . No explosions," said Lieutenant Wharton from *PT-508*.

"Request permission to proceed independently," said Crist. "I want to get in closer."

"No heroics," said Sherertz. "Surprise 'em, spray 'em, and split."

Over the radio came the sound of three Packard engines wide open, followed immediately by gunfire. In seconds, they could hear gunfire with a different sound, originating from the German ship.

"We're right in the middle of them," said Crist.

"We can't see a thing," said Wharton. "Both sides are still firing. We can't fire or we might hit *Lobster*. *Lobster* has disappeared off our radar screen. Do you see them?"

"Just one blip, near you," said Scott. "Can you raise

them?"

"No, but they're both still firing," said Wharton.

Boom!

"What was that?" said Sherertz. "*PT-509*, do you copy? Report in, please. Lieutenant Whorton, what happened?"

"It's suddenly dead quiet," said Wharton. "Report in, *509* . . . Nothing, sir."

"All boats return immediately," said Sherertz. "Out."

"What's your plan?" asked Scott.

"Scotty, I think they sank her," said Sherertz. "I'm going out looking for survivors."

~~~

At 0725, Lieutenant Sherertz stood on the bow deck of PT-503 with its assembled men, adjacent to *PT-507* with its men.

"This is a rescue mission," he said in a loud command voice. "Anybody who thinks he's going for revenge can get off right now. Any questions?" He paused for a moment. "Everyone have his flak jackets on? All right, then, move out!"

Sherertz felt quite apprehensive about what they would find. He could think of no good reason why the *509* boat would both disappear off the radar screen and stop communicating.

~~~

"You're on vector," said Scott over the radio. "Straight

347

as she goes. Target three miles. Target two miles. Target one mile. Five hundred yards."

Strangely, the target was two, nonmoving radar blips.

"Go straight at them, dead quiet," said Sherertz to Ensign Schuster at the helm. "If it's the *509*, we'll stop to search for survivors. If it's the Jerries, we'll fire our remaining torpedo and then turn hard to port and open up with all guns. Stay on our starboard wake, *507*, fifty yards back, and follow our lead."

"Lieutenant Sherertz, if we get much closer, we're going to ram," said Radioman Kramer from the chart house.

Figure 103. RON 34 PT boats moving in fog.
Jack Sherertz collection.

As the *503* moved slowly forward, the fog suddenly cleared. Less than one hundred yards dead ahead was a German minesweeper, almost three times the size of a PT boat. What followed seemed to occur in slow motion. "Doc" Schuster was at the helm. To his right was RON 34 CO Lieutenant Sherertz, with Ensign Koenen behind them on the left and PT-503 boat captain Lieutenant Jim Doherty to Koenen's right. Gunner's Mate Sullivan manned the forward 37 mm gun, with "Duke" Duquette supplying ammo. Motor Mech Allbee was on the forward starboard .50-cal machine gun, and Motor Mech Albright manned the aft port .50 cal. Gunner's Mate Brumm trained the aft 40 mm gun, with Radioman Fisher controlling elevation, Gunner's Mate Biele loading, and Quartermaster Lang passing ammo. Below deck was Radioman Kramer in the chart house and Motor Mech Peppel in the engine room.

"Doc, come right two degrees and gun it," yelled Sherertz. "Duke, fire the torpedo."

"Open fire!" yelled Lieutenant Doherty and Ensign Koenen.

Figure 104. Forty mm cannon at rear of PT, firing away.
Jack Sherertz collection.

PT-503 fired all guns, followed seconds later by return fire from the Germans.

Tat, tat, tat, tat, tat, went the .50 cal, but suddenly everything seemed to slow the down for Sherertz. *Tat . . . tat . . . tat . . . tat . . . tat . . .*

He felt he could see the impact of each shell. One German sailor aiming a rifle was hit in the chest and blood sprayed out behind him. Others adjacent to him dove for cover. More men were pouring out from below deck to man the guns where others had fallen.

Boom went the 20 mm and *Booom* went the 40 mm. The Germans responded with similar caliber return fire.

The sound was deafening and getting louder by the second as the distance closed between the ships. A high tracer-to-bullet ratio in the .50-cal ammo belts caused the PT fire to look like a laser light show, lighting up the fog a brilliant white.

Sherertz could see continued damage by their 50 cals, but he also saw German return fire tracked these lines of light with deadly accuracy.

Booom came the sound of a 37 mm German cannon, sending a shell shrieking over PT- 503's bridge.

Kaboom went an exploding 37 mm shell, striking the aft .50-cal turret. Shrapnel exploded into "Pop" Albright's chest, knocking him out of the chair. He slumped over as he landed and disappeared into the turret cylinder.

Kaboom went another exploding shell, this one striking

the radar mast and showering everyone near the bridge with shrapnel. The back of Doherty's helmet took a piece of shrapnel that went through his scalp, snapping his head forward. He reached back to find blood pouring down his neck. Allbee, in the forward .50-cal seat, and Sherertz were both hit, but neither of them flinched.

"Do you have engines?" Sherertz asked Doc.

Schuster glanced at all three tachometers. "Yes, I have power."

"Let's get the hell out of here," said Sherertz. *They're going to tear us apart if we don't.* "Hard to port."

PT- 503 banked hard, coming within ten feet of the German minesweeper, close enough for Sherertz to see the German's faces.

Gunner's Mate Brumm on the aft 40 mm was firing as fast as Biel and Lang could reload.

Booom . . . booom . . . booom . . . booom went one shell after another, causing major damage to the minesweeper. The Germans now trained their guns on Brumm. Suddenly he slumped to the deck, shot in the chest. He did not move after that.

Kaboom came the sound of another shell hitting the smoke generator in the stern of the boat. Smoke began pouring out. *PT-507* on the starboard wake of the *503* was also firing all guns, but the smoke from the *503* obstructed their vision, so they ceased fire.

"*507*, get your ass back to the *Maloy* on the double,"

yelled Sherertz. "Tell Scotty we have wounded."

Sherertz and the other officers went looking for the injured. They found Brumm, still alive but unconscious, and moved him to the portside bridge area. Setting him down, they attempted to stop the bleeding, but it was no use. Blood poured out of the large wound in his chest.

Biele, Fisher, Allbee, Faucher, Peppel, Lane, and Sherertz were all wounded, but fully functional.

"Anyone seen Albright?" asked Doherty.

"Oh damn," said Doc. "He's in the .50-cal well."

They pulled him out and tried to stop his bleeding. He had a pulse, but just barely. When they got back to the DE *Maloy*, they moved Brumm and Albright over to the *Maloy* and into sick bay. Brumm was already dead. The chief pharmacist's mate had been alerted and was ready to administer to Albright. Sherertz went with them.

~~~

Sherertz sat down next to Brumm, on the side of the gurney opposite the pharmacist's mate. Brumm's uniform had been cut back, revealing a large wound right over his sternum. Lieutenant Sherertz took Brumm's hand.

"Gunny, you did a great job back there," said Sherertz. "We're all proud of you. I plan on writing your family and telling them all about what you did. Don't worry about anything. They're going to take good care of you. Why don't we say a prayer? Dear God, Gunny Brumm needs your help. Please take care of him and guide him on his journey. Amen."

Sherertz then reached over, grabbed a clean bandage, and dipped it into a bowl of saline. Meticulously, he wiped the blood off Brumm's face. He wanted to make sure he didn't miss a single drop. Finally, he looked over to see how the pharmacist's mate was doing with Albright. The medic had stopped working.

"He's stable for the moment, sir," said the medic, "but I don't think he's going to make it. Would you like some time alone with them?"

Sherertz nodded. He finished cleaning the blood off Brumm's face and neck, pulled Brumm's shirt back over his wound, and crossed Brumm's arms over his chest. Then he placed his own hands across Brumm's hands and looked at him, tears streaming down his face. After a while he stood up, reached for some unused gauze and the bowl of saline, and followed the same procedure with Albright. Finally, he took one last look, saluted, and walked out.

~~~

August 11, Cherbourg Naval Base

On the parade grounds near the Green Barracks on the west side of Napoleon III basin, in front of a flagpole flying the US flag, were Lieutenant Sherertz, RON 34 XO Daniel, and a chaplain. To their right was a seaman with a bugle. Sherertz nodded toward him, and he played the call to assemble. All of RON 34 and the base and tender staff were ready for the memorial service and came running out to the parade grounds double time in their dress blues. They spread out in four rows of about sixty men, each standing at attention.

"At ease, men," said Sherertz. "You all know why we're here. They sank *PT-509*, with all men presumed lost. In an attempt to try to rescue them, we lost two more men from the *503* and have two more at the base hospital. All of our dead gave their lives for our country. They gave their lives for us. I have served with many men since the start of this war, and none finer than the ones we lost today. All of you knew them. I want you to take a moment of silence and think of the good things you know about them and say a prayer that they are safe and out of danger. After that, I would like the chaplain to lead us in the Shepherd's Prayer."

They all bowed their heads and observed a moment of silence.

"The LORD is my shepherd. I shall not want," began the chaplain. "He maketh me to lie down in green pastures: he leadeth me beside the still waters. He restoreth my soul: he leadeth me in the paths of righteousness for his name's sake. Yea, though I walk through the valley of the shadow of death, I will fear no evil: for thou art with me. Thy rod and thy staff they comfort me. Thou preparest a table before me in the presence of mine enemies: thou anointest my head with oil. My cup runneth over. Surely goodness and mercy shall follow me all the days of my life. And I will dwell in the house of the LORD for ever."

"XO, please call the roll of the men we lost today," said Sherertz.

"PT- 509:

Gunner's Mate Third Class Ausley

Radarman Third Class Bricker

Motor Mechanic Second Class Horsfield

Gunner's Mate Third Class Kornak

Ship's Cook Third Class Line

Motor Mechanic Second Class Lossin

Radarman Third Class Page

Radioman Third Class Reynolds

Gunner's Mate Third Class Ricci

Torpedoman Second Class Schaffroth

Quartermaster Third Class Thale

Gunner's Mate Third Class Wypick

Lieutenant Junior Grade Mathes, Third Officer

Lieutenant Junior Grade Pavlis, Executive Officer

Lieutenant Crist, Commanding Officer

PT- 503:

Gunner's Mate Third Class Brumm

Motor Mechanic Second Class Albright"

"Chaplain," said Sherertz, "will you say a few words?"

"Men, nothing I can say can stop the hurting you feel inside, but know this. All of these men were God-fearing souls, and I strongly believe they have arrived in heaven. We must continue what they started, as they would have wanted us to do. Let us pray."

Everyone bowed their heads and remained silent for several minutes.

"Men, I will be writing a letter to each of these men's families," said Sherertz. "If you have anything you would like me to include, please bring it to my quarters by twenty-one hundred hours tonight. Chaplain, please lead us in a closing prayer."

"Dear Lord, blessed are those who sacrifice their lives for the good of others. Please guide them in their new life and let them know they will be remembered."

"Squadron, attention. Dismissed," said Sherertz. As he watched, the men slowly began to disperse.

Boom came the sound of a distant bomb exploding, and just that quickly, Jack saw their faces—Jones, Christopher, Sterns, Davis, Booth, Uhrenholdt. He heard the sounds and smelled the smoke of burning oil and the unforgettable odor of burning flesh. Then the faces of Brumm and Albright, and the German minesweeper, with all its guns shooting at him. As the memories flooded in, he broke out in a sweat, and his knees felt rubbery. He took a deep breath and stiffened his body to keep from falling. He must get back to his quarters. With sheer force of will, he blocked everything out and began walking, his muscles rigid. He walked through the men, not noticing them. Ensign Russ Schuster and Ensign Stan Allen had to step back to avoid a collision.

They stared after him as he walked toward the barracks. "What's up with Ramrod?" asked Allen. "His face looked horrible."

"I think he's trying to deal with sending sixteen men to their death," said Schuster. "And now he's got to figure out what to say to their families."

"And in front of all of us, act like it's not bothering him," said Allen. "I'm glad I'm not in his shoes."

Figure 105. Only part of *PT-509* found after it disappeared.
Jack Sherertz collection.

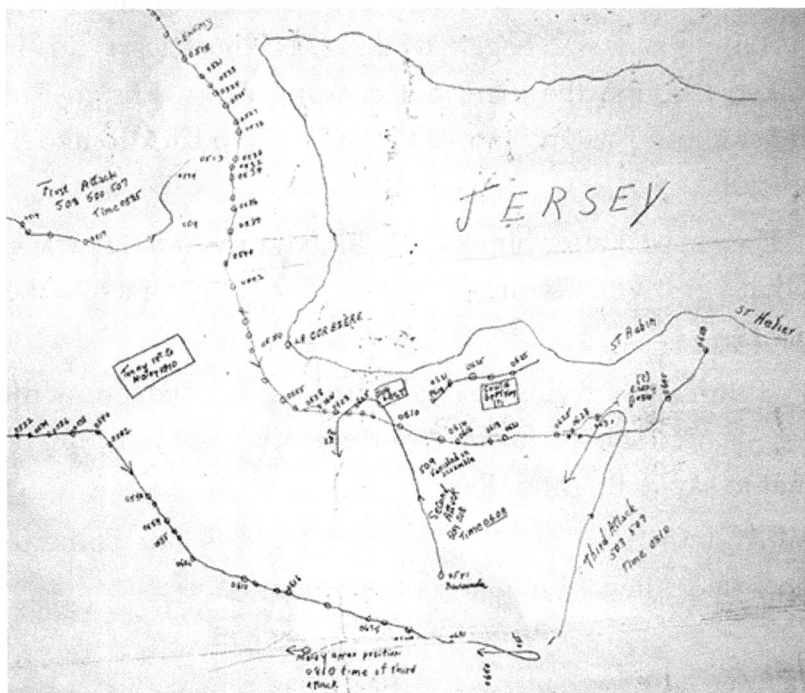

Figure 106. Path of German convoy (10 boats) around Jersey Island on night of August 9 1944. RON 34 first attack 05:35, approximately 2 miles due West. RON 34 second attack 06:08, less than 1 mile southeast of La Corbiere. RON 34 third attack (rescue attempt) 08:10, a few hundred yards south of Noirmont Point. John Ovenden collection (see acknowledgements).

<u>German Communique August 10th</u>

A large enemy destroyer was damaged badly by a bomb hit in waters
west of Brest. The enemy once again
lost six fully-laden supply vessels
totaling 25,500 tons, one destroyer
and one escort vessel in the Seine Bay
through special weapons of the German
Navy. Vessels escorting a German convoy
sank two American E-boats south of Jersey
Island, one by ramming in close-quarter
fighting.

<u>ALLIED COMMUNIQUE August 10th</u>

PT-boats on an offensive patrol off the
island of Jersey intercepted a southbound
enemy convoy off Corbriere in low visibility
yesterday morning. One enemy vessel was hit
by a torpedo and one other was damaged by
gunfire before the convoy escaped. Shortly
after daylight, with visibility further reduced
by fog, other PT-boats entered the roadstead of
St. Helier and attacked two M-class minesweepers
with gunfire. Many hits were observed on the
enemy before our forces withdrew.
Off the port of Le Havre light coastal forces
intercepted an enemy vessel under the escort of
six R-boats. The enemy vessel was sunk. One of the
escorts also was hit by a torpedo and its destruction was considered probable.

Figure 107. Copies of US, German communiqués issued after *PT-509*
disappearance. Jack Sherertz collection.

World War II Final Assignments

XO Daniel came up to the Cherbourg dock where Lieutenant Sherertz was doing an inspection.

"XO, I can't believe the Navy is going to make us turn over RON 34 to the Russians," said Sherertz. "Why not give the boats to France to help them rebuild, or anyone else, but not to Stalin. Next thing you know he'll be shooting at us from our own boats. It doesn't make any sense."

"I agree, Skipper, but we don't have any stars, so we don't matter. Oh, I almost forgot. I just got a message from Sparks. Captain Bulkeley's in that launch you see coming into the dock. He wants to talk with you."

~~~

"Lieutenant Sherertz, I heard Ike is going to give your squadron to the Ruskies," said Captain Bulkeley. "That's a really bad idea, on so many levels. I believe that leaves you without a command. His folly may be my advantage. I've come with an offer. I want you to be my XO on the USS *Endicott*. They just shipped my XO to a different ship, and I think you're the best man for the job. I know you were just approved for Lieutenant Commander. If you come my way, you'll receive an additional promotion to Commander. If you do the great job that I know you will, within a year you'll be promoted to captain of a destroyer or DE escort. I think that's a pretty sweet deal. What do you think?"

Sherertz looked hard at Bulkeley, then out toward the water, and then back at Bulkeley. "Skipper," he said, "it's a great offer. No, it's a fantastic offer. A few months ago, I might have taken you up on it, but I see my time in the Navy coming to an end. I've found a job as an experimental officer back at Warwick. They're going to let me test all of the new equipment until the war is over. Then I'm becoming a civilian."

"That will be such a waste of ability," said Bulkeley. "There are so few officers that can lead men, who the men actually respect. You could be a star in the post-war Navy, Sherertz."

"All the same, John, I've made up my mind. Thank you for coming in person." Sherertz extended his hand

Bulkeley hesitated, looking hard at Sherertz. He didn't flinch; his face was rigid. It was clear he wasn't going to change his mind, so he shook Sherertz's hand.

Sherertz couldn't bring himself to tell Bulkeley the truth. As good a relationship as he felt he had with him, he didn't think "jump on the Japanese barge with guns blazing" Medal of Honor winner Bulkeley would understand. It wasn't that Sherertz was afraid. It just wasn't in his nature to send men that he trained, men that he loved, out into harm's way and have them come back dead. One more round of that, and he might go to some dark place and never come back again.

"Don't be surprised if you hear from me again," said Bulkeley. "I don't give up easily, once I've spotted someone with talent."

~~~

Figure 108. Lt. Sherertz with Russian officers.
Jack Sherertz collection.

December 1944, Rosneath, Scotland

Lieutenant Commander Sherertz arrived at the docks to talk with assembled RON 34 officers. He had been ordered by Allied High Command to turn over all of his PT boats to the Russians under the recently passed Lend-Lease act. It was passed by the US Congress in March 1941 and resulted in the transfer of food, oil, and other materials valuable in the wartime effort to Great Britain, France, China, and the USSR. Neither Sherertz nor his men were happy about it.

"Gentlemen, I just finished meeting with Captain Kozlov and Lieutenant Commander Baudarivk about our ten PTs," said Sherertz. "The damn Ruskies expect that all original Elco equipment and supplies will be on each boat. I know all of you did your best to scrounge up everything that was missing, but we know we're short. However, I think we can pull this off." He leaned in conspiratorially, and they all huddled, as if in a US football game.

Moments later, the two Russian officers arrived to begin the inspection process. Sherertz, XO Daniel, and the CO of each boat were allowed to accompany them on each inspection.

As they went through the first boat, everything seemed to go well. The Russian officers seemed especially pleased when they got to the galley.

"Ahga diir iz ze pikkl dish," said Kozlov.

As they climbed up on deck, Sherertz and Daniel made an elaborate show of wanting to examine the list to be sure that they had checked off everything. Meanwhile, the boat CO had walked behind the bridge and dropped a carefully wrapped package over the side to the men on the out-of-sight rowboat waiting below. While the diversion was going on above, they rowed to the next PT, and the missing items were transferred on board so that it was ready to be inspected. Meanwhile, the inspection of the first boat was concluded.

"Verrie goot , ol iz in orrderr," said Baudarivk. "Aur man vill sil up zis bot and moove to ze nekst uan."

In short order, all ten boats were signed, sealed, and delivered.

~~~

March 13, 1945, Melville, Rhode Island

"Welcome back, Sherertz," said Johnny Mote. "It hasn't been the same since you left."

"Good to be back, old friend," said Sherertz. "I'm kind of excited about this job as experimental officer. It will give me a chance to fix all of the things we kept complaining about in the field. They tell me I'll get to test out *PT-796*. She's a Higgins boat with mahogany on the outside and spruce on the inside, and similar armament except for one very important difference. It has two Mark 50 5-inch rocket launchers. That should be fun."

"Yeh, maybe so, but why did you turn down a plum job as Captain Bulkeley's XO on the USS *Endicott*," said Mote.

"Who did you hear that from?" Sherertz shook his head at the speed of the naval grapevine.

"You know that would have put you on the fast track to naval greatness," said Mote with a grin. "You would have made captain within a year and then been able to move up to one of those big boats you wanted at the beginning of the war."

"Naval command isn't for me," said Sherertz with a distant look. "And you were right about the ass-kissing. I think I'd rather stick around here with you and do something creative."

"What about medical school? Didn't you tell me you wanted to be a doctor?"

"Yeah, that was the original plan. But I think I've seen enough death and dying to last a lifetime."

~~~

May 8, 1945, Melville, Rhode Island

Sherertz was up on a docked PT boat with a bunch of new PT boaters, explaining the basics of a .50-cal machine gun when Johnny Mote ran up on the dock.

"Germany surrendered!" yelled Mote.

Everyone cheered and started jumping up and down, but Sherertz just stood there, feeling an overwhelming sense of relief.

"Sherertz, there's more. I think you may even like this better. Here's a telegram from J. J. Daniels marked *urgent*. My gut tells me it's something important."

Sherertz grinned. "Your gut or what you read holding it up to the light?"

"Ah, just read the telegram, Sherertz," said Mote.

"I'll be right back, gentlemen," said Sherertz. "I expect you to be able to demonstrate what I just showed you when I get back." He ran over to the edge of the boat and jumped down to join Mote. "Let me see that." He opened up the telegram and took a minute to read. "Unbelievable! Page is alive! Did you hear me, Mote? Page is alive! He's been a prisoner of war since the *509* sank, and the Germans just now let him go."

"Page is alive, Page is alive!" shouted Mote. With tears in their eyes, he and Sherertz danced around on the dock like a couple of crazies. Through the tears, they couldn't stop grinning at each other and slapping each other on the back.

~~~

Figure 109. USS *Acontius*, Leyte, Philippines.
Jack Sherertz collection.

August 15, 1945, Leyte, Philippines

In spite of protests about having to leave Warwick, Rhode Island, Lieutenant Commander Sherertz had been transferred to the USS *Acontius* under the command of Commodore Richard Bates to become the PT Pacific fleet supply officer. Today he was sitting at a table in the officers' mess having lunch with Commodore Bates and a group of rookie officers.

"I wish I was there when they dropped Little Boy on Hiroshima on August 9," said one officer. "That must have been quite a show."

"I heard that the Fat Man bomb they dropped on Nagasaki was even bigger," said another, in a one-upmanship attempt.

Sherertz winced inside at their lack of understanding at how many lives must have been lost with each bomb.

"What are you going to do after the war, Sherertz?" asked one of the other officers.

Sherertz was about to answer when a young officer came rushing in, holding a piece of paper in his fist. He handed the crumpled paper to Commodore Bates, who read it quickly.

"The war's over, men," said Commodore Bates. He seemed to take it in stride.

At last, Sherertz thought to himself.

The young communications officer went running out, yelling the news to anyone and everyone. The rest of the men in the room began whooping and hollering, and a general celebration began that went all night.

Figure 110. Celebration of V-J Day, Leyte, Philippines, harbor.
Jack Sherertz collection.

~~~

"Lieutenant Commander Sherertz, Commodore Bates will see you now," said Bates's assistant as he came out of Bates's office. Sherertz went through the open door and closed it behind him.

"At ease, Sherertz," said Bates. "I have one more duty for you before you go stateside."

Sherertz was not happy to hear that. He had thought that he was going to get his orders to head home. This seemed entirely too formal, given that the war was over. Bates seemed to be struggling to find the right words, so Sherertz just stood there and waited.

"There's no good way to put this, so I'll just spit it out," said Bates. "The order has come down from High Command to burn all the Pacific PT boats, and I'm asking you to oversee that order."

"You're kidding, right? First they make me give my RON 34 boats to the Russians, and now you want the boats that were at Pearl Harbor, Midway, and the rest of the Pacific Islands to be burned. This is not funny at all. In fact, it's sacrilegious."

"It's not a joke, Sherertz," said Bates. "Here's the order."

Sherertz reached out and took the paper from Bates's hand. Reading it quickly, he looked up, searing anger gripping his face. "And there's no one else you could find to carry this out?" he said in a controlled, tight-lipped voice.

"I think you're the best man for the job," said Bates.

"You just don't get it, do you?" Sherertz asked, his voice edged with anger. He shook his head. "Is that all, sir?"

"Well, I thought we might have a drink together before you left."

"No, thank you. You've made me lose my appetite for just about everything." Sherertz glared at Bates.

"Well, in that case, you're dismissed."

~~~

Lieutenant Commander Sherertz stood on a beach at PT base 17, at Samar Philippine Islands near Bobon Point, staring at the fleet of PT boats anchored in a shallow cove. In front of him a few hundred feet off shore were tied up more than 100 PT boats. They had been stripped of armament and engines, naked hulls, ready to be burned. He could barely bring himself to look at them, knowing what was about to happen.

He addressed the Marine sergeant next to him. "Sergeant, as you can see, this is going to be quite a bonfire, which I would rather not see. Are you still okay with overseeing this?"

"Yes, sir," said the sergeant.

"Thank you," said Sherertz, shaking his hand. "I am eternally grateful. I leave things in your capable hands, but wait until I'm out of sight. I don't want to see or hear anything of what's about to happen."

As Sherertz turned to leave, he suddenly realized this was his last military action. He had wondered previously if some deep meaningful thought might occur at this moment, but all he could think about was when was the next flight out of Manila. Without further ado, he walked away and didn't look back.

Figure 111. Samar PT Base 17 at Bobar Point, Philippines.
www.ibiblio.org/hyperwar/USN/CloseQuarters/img/PT-p419.jpg

Figure 112. Burning of Pacific fleet PT boats at Samar PT Base. Bill Maloney
photograph. Image can be found at http://www.williammaloney.com/dad/
wwii/MiltWWII/WarsEnd/pages/03BurningTheHulls.htm

or at www.AHeroAmongMillions.com.

Part III

Final Tributes

Family Recognition

The month after the 2009 D-Day celebration, I took my father and his wife, Lillian, to the next to the last PT Boat Reunion in Arlington, VA. Upon arrival my dad and I went down to the reunion area of the hotel and the first person we saw was a man my father's age, in a wheelchair, being pushed by a man my age. We walked up to them. My father looked at the older man and he looked back. Initially there was no evidence of recognition.

"My name is Jack Sherertz, and this is my son, Bob," said my dad.

I'll never forget the responses on the face of the man in the wheelchair. First a look of recognition when he heard my Dad's name, then visible joy with tears. When it was clear that he was unable to speak, his son introduced himself and said, "This is my father, Captain Earl Fox."

That introduction allowed Earl Fox to recover. "I was in RON One at Pearl Harbor," he said. "After the Nevada was sunk by the Japanese, your father was reassigned to RON One. His first assignment was to *PT-22*, the *Flying Deuces*, my boat. I was the XO. The only position left open in the squadron was a doctor's slot, which we gave to Jack." "What we really needed was someone who knew about guns to train our gunner's mates, and after sevens months on the *Nevada*, Jack knew more about guns than almost everyone in our squadron."

My dad listened intensely with tears streaming down his face. Finally, I motioned to Earl's son, Parham Fox, and we stepped to the side. My dad and Earl talked for a few minutes and then grasped each other's hands, followed by more tears and finally parted company without saying another word. We were fortunate enough to also meet a few men from RON 34 with the result being that my dad had decided he wanted to return for the last reunion.

~~~

In December 2009 my family held a ninety-second birthday party for my father that I organized to recognize his World War II military service. I put out all his World War II artifacts, so he could talk about them, and then we gave him cake and ice cream. At the end we gave him three very special gifts. The first was a wooden model boat that he had purchased during World War II that I arranged to have built by a model builder named Alex Johnson. It had painted on it *PT-22, Flying Deuces*. The second gift was a shadow box. At the top was the headline from an extra edition of the *Roanoke Times*, stating, *U.S., JAPANESE AT WAR*. Around the sides were his Navy rating stripes, rating pins, a nameplate that said, *Lt. Commander Herbert J. Sherertz, Supply Officer*, a summary of his World War II military service by date, and some other pictures. In the middle was his Bronze Star, Purple Heart, and British Distinguished Service Cross, and below that his service ribbons. Finally, lower still, was his dog tag. The third and final gift was a bound copy of the first version of his story, then titled, 'One Hero Among Millions'. By the end he was reduced to tears and unable to speak; he clearly appreciated our efforts.

Figure 113. One Hero Among Millions. R. Sherertz collection, taken by Eric Sherertz and included with permission.

~~~

In April 2010, I was able to take my father and two other veterans (Staff Sergeant Curtis Winfred 'Wimp' Wise and Aviation Metalsmith 2nd Class Raymond Van Havere) on a North Carolina Rotary sponsored World War II Fflight of Hhonor. He wore the same Navy Blue suit with the ribbons and medals on it that he had worn to the 2009 D-Day reunion. On the Flight of Honorthat trip I learned several important things about World War II veterans. First, they go to the bathroom a lot. Second, they like being recognized for their service. After thirty minutes of most of the veterans being ignored by the tourists at the World War II memorial in Washington, DC, while my father was being mobbed by tourists who noticed his ribbons and medals, a group of his fellow veterans came up to him and wanted to borrow his coat. And third, my father always told me that he had never met a stranger. Proof of that statement could not have been more evident than at the receiving lines at the Reagan National Airport and Charlotte Airport when he shook the hand and spoke to everyone in the receiving line, and I do mean everyone.

~~~

In July 2010 my two sisters, Lillian, and my dad arrived at the hotel in Warwick RI for the last PT boat reunion in July 2010. Lillian was tired and wanted to rest. Not my dad. He immediately headed down to the reunion. Downstairs, we received our nametags from Alice Guthrie, the daughter of Jimmie "Boats" Newberry, the founder of PT Boats, Inc. My sisters were amused to find out that their name tags and mine all had the word *Splinter* on them, as in splinter off the wooden PT boats, or descendants of a World War II PT boater.

We could have gone to the site of the World War II PT boat training facility, but none of the original buildings were there and I felt we needed to conserve Dad's energy, so I made the executive decision to head to the PT Boats, Inc. store area instead. It was quickly apparent that many PT boaters had made the same decision. As my dad struck up a conversation with one of them, my sisters caught up to me.

"Yo, brother, what's the story about this place," said my sister, Sue, having just arrived with my other sister, Meg.

"So sisters, this is the PT Boats, Inc. store," I said. "PT Boats, Inc. is a nonprofit devoted to preserving the records of PT boats and PT boaters from World War II. Hundreds, if not thousands, of World War II PT boaters have donated records, pictures, and artifacts that now reside in two locations: Germantown, Tennessee, and Battleship Cove in Fall River, Massachusetts. The former houses all of the men's war records, pictures, artifacts, etc. The latter houses two restored PT boats. The latter is the reason I wanted to come to this reunion, to get on a PT boat with Dad and see it through his eyes." A quick glance at myMy sisters told me they werelooked bored, so I cut it short. "Would you like to wander around the store and shop?"

"Now you're talking," they replied together. And so we shopped.

That night we had dinner with some PT boaters and their families that I hadn't met previously, and shortly thereafter we went to bed, all tired from a busy, busy day.

~~~

The next morning we ran into a man named Danny Diaz, a young film producer making a documentary about the last PT boat reunion. I took Dad up to the room where they were interviewing the PT boaters and stood to the side as they talked to him. Dad told a few stories and then it was time to get on the bus to head to Battleship Cove.

When we got there, some people went to tour the USS *Massachusetts*, but I shepherded our group straight to the PT boat museum. Inside the museum, Lillian stayed in her wheelchair, but not Dad.

"Bob, which boat do you want to see first?" he asked with a twinkle in his eye.

And off we went. It looked as if the Elco boat had most of the people on it or around it, so we headed to the Higgins boat first. Usually tourists are only allowed to walk around the outside of the boats and look through windows cut for viewing. However, the PT boaters and their families were allowed to go anywhere, including inside the boats.

"Say, Dad, were these boat hulls really made out of plywood?" I asked.

"No, Bob, they were made of two one-inch layers of mahogany, with a layer of canvas in between."

As we walked around the outside walkway, he

frequently stopped and stared inside for a few minutes, again with that distant look. We waited until he was ready to move on.

"Bob, let's go up on deck."

We climbed up onto the boat's stern (rear end for landlubbers), and he took us over to see a Mark VIII torpedo, the namesake of PT boats (Patrol Torpedo Boats). The original boats only had one on each side. Later in the war, they carried two per side. The torpedo we were looking at, sitting in its housing, was much harder to launch because it essentially had to roll off a ramp rather than be shot out of a tube with compressed air.

"Impressive, aren't they?" Dad said. "They could travel fourteen miles at thirty-six knots and carried over four hundred pounds of TNT. But they couldn't hit the broad side of a barn. Neither squadron I served with hit a single ship."

Next he took us over to the 40 mm canon on the back of the boat.

"What's this gun for?" Sue asked.

"We used the 40 mm cannon to shoot at other boats, enemies on the shore, or low-flying planes." Dad grinned and pointed to the stern hatch. "Want to go down below?" he asked in an animated voice.

Figure 114. Jack Sherertz about to climb down inside PT boat.
R. Sherertz collection.

I went down the hatch first so I could help him down, or break his fall if necessary. With some support from below, he did fine.

"What's the desk for, Dad?" Meg asked.

"It was supposed to be used by one of the officers to plot the course or do paperwork. In actual fact, I spent very little time at such desks because there was always something that needed to be overseen on deck. Plus, it was very difficult to write when the boats were underway, bouncing over the waves."

Figure 115. PT Boat Officers Desk.
R. Sherertz collection.

"How about a picture so we can see how you might

have looked?" I said, snapping one before he could answer.

"Where's the kitchen?" Sue asked.

"You mean the galley," I corrected with a grin.

"Yeah, where's the galley?" Sue grinned back.

"You're looking at it," said Dad, pointing behind us at a hot plate on a counter. "We roughed it."

My sisters and I laughed.

Next to the food preparation counter was a small table with benches and a bunk area built into the space above the table, next to the boat's hull. PT boats wasted no space.

We took a few more pictures and finally came up the bow ladder to the front of the boat. No big person would be able to move up that ladder or out of the bow hatch, so overweight sailors could not have served on a PT boat.

I could see that things had cleared on the Elco boat, so we headed that way. The tour down inside the Higgins boat had winded my dad, and he was moving slower and seemed out of breath. When we had finally ascended the stairs leading up to the deck, I paused to give him a little respite. But he didn't stop. Instead he went straight to the bridge.

"I spent a lot of time up here during the war," he said as he held on to the boat's steering wheel. He looked at home, comfortable behind that wheel. He got quiet and stared off into space for a few moments, dredging up more memories.

Facing the bridge, there was a .50-cal machine gun

turret to the right and to the left, with the radar equipment directly behind and the signal light to the right of the radar stand. My dad's head barely peered over the top of the bridge, whereas seventy years previously, he stood almost a foot higher and, and based on pictures from his World War II albums, he had a commanding presence.

Figure 116. Lt. Commander Sherertz and daughters on the bridge of a PT boat. R. Sherertz collection.

"Let me show you a few things about how my boat worked. There were three throttles that operated independently of each other . . ." He told us how he and his men estimated speed, how they slowed down in a hurry. In short order, my sisters' eyes started to glaze over. I knew I needed to keep things moving. We closed the tour of the PT boat with a group picture.

Figure 117. Jack Sherertz and children on bow of PT boat at Battleship Cove, Massachusetts. L-R: Susan Anthony, Jack Sherertz, Meg Pearce, Bob Sherertz. R. Sherertz collection.

We went back and picked up Lillian and headed to the USS *Massachusetts*. When we got to the ramp to go up on deck, there was some brief head scratching regarding how we were going to get Lillian and her wheelchair up the ramp, but it quickly became a nonproblem. Four active Navy men picked up her wheelchair fore and aft and walked up the ramp with her. As soon as we arrived on deck, we found seats for Dad and Lillian down front, near the podium and band, and seats for my sisters and me farther back. We hadn't been there long when I saw Lillian motioning for me to come. When I got there, it was clear that my dad was not doing very well.

"What's wrong, Dad?" I asked.

"My chest hurts," he said. He was quite sweaty, much more so than anyone around him. I spotted a Navy medic and told him what was wrong. He grabbed a wheelchair and helped me get Dad into it, and we took him inside, where it was air-conditioned. His blood pressure was initially a little low, and his pulse was on the slow side. In only a few minutes, his chest pain was gone, and his blood pressure and pulse were back to normal, but we decided to let him cool down further before taking him outside in the heat again. It was clear that my dad was pushing hard. I think he increasingly felt his time was running out.

Finally, with my dad doing much better, we went outside for the roll call ceremony for those PT boaters who had died in the past year.

Figure 118. 2010 roll call of all PT boaters who have died in the previous year, deck of the USS *Massachusetts* at the last PT boat reunion.
R. Sherertz collection.

One by one, more than two hundred names were called out, followed by a stirring twenty-one gun salute fired from the five-inch guns on the USS *Massachusetts*. Only a few hundred PT boaters were now left out of the more than seventeen thousand who had participated during World War II.

There remained only the dinner that evening, and the last PT boat reunion would be finished.

As we walked toward our dinner table my dad was fortunate enough to run into the current Commander of the modern-day PT boats known as Special Warfare Combat Crafts (SWCC) or Fast Boats, for short. Nowadays, these boats transport and support special forces operations. He and my dad had a nice conversation.

At dinner, my dad was quite animated. He seemed more interested in talking than eating, though at times I wasn't sure whether he was talking to us or merely remembering things out loud. No doubt the events of the day had stirred quite a few memories.

"You know I outranked President Kennedy," he said with a grin, telling me something he had mentioned several times previously.

When we had all finished eating, I asked my dad if he wanted to talk to anyone in particular, as this might be his last chance. He said he would like to talk to some of his squadron mates. I went looking and found three of them: Russ Schuster, Stan Allen, and Norm Hoyt. They were all busily talking with family, but said they could be free in about an hour, if that were possible.

Figure 119. RON 34 squadron mates (left to right) Norm Hoyt, Stan Allen,
Jack Sherertz, and Russ Schuster at the last PT boat reunion.
R. Sherertz collection.

When my dad's squadron mates joined us, they began talking about World War II.

"Can you believe it?" asked Norm Hoyt. "The Navy trained me six weeks so I could cook for hundreds of men, when all I needed to do was make sandwiches and heat up cans on a hot plate. What a bunch of idiots."

Stan Allen told the story of how they almost blew the D-Day surprise attack on June 4, 1944, by attacking the British destroyer HMS *Saumarez* that was sent to call them back because of bad weather. In a spot-on imitation of a British accent, he recounted what the British officer had said.

Finally we heard Russ Schuster talk about what it was like driving *PT-503* as our dad yelled out commands in the battle with the German minesweeper.

As I watched my sisters listening to the stories, I realized that they were finally beginning to understand something about what he experienced during the war. Mission accomplished.

The next morning, the last PT boat reunion was over. As we waited for a taxi, I struck up a conversation with a PT boater named Clyde Combs (*PT-515*, Squadron 35). He had also been part of the PT boat Channel operations after the D-Day assault. He told me about going to Normandy to receive the French Legion of Honor Medal from Nicolas Sarkozy during the 65th D-Day memorial celebration that President Obama and Prince Charles attended. When I asked him what he did that made him eligible for the medal, he replied that he did the same things my dad had done, and that my dad was eligible.

Figure 120. Completed model PT boat painted to resemble Lt. Cmdr. Sherertz's first Patrol Torpedo boat (PT- 22), the *Flying Deuces*, that he joined up with in December 1941 after the Pearl Harbor attack. R. Sherertz collection.

D-Day, 2012

In the midst of all of the events we had organized for my dad I received an invitation to give a talk in Lille, France, on June 6, 2012, D-Day. At that time my father was still healthy and able to walk, so hoping that we could both participate in the French D-Day celebration, I agreed to come. Unfortunately my father suffered a stroke in January 2012 and nearly died three times between then and June. Now in France, a few days before D-Day, I was quite anxious about whether he would live until I returned home.

I tried to put that out of my mind as my daughter and I prepared to make an all night drive from Paris to Saint.-Malo and then on to Jersey Island by Ferry. Shortly after leaving Paris city limits, my daughter fell asleep, but I knew the 6-20 ounce caffeinated sodas at my side would keep me awake. Early in the trip I saw one road sign after another warning about deer, I tried to use the high beam headlights, but oncoming traffic required frequent dimming. Just as I had given up on using high beams, my dad, always a stickler for safety, popped into my head telling me to turn on my high beams. I did, and there fifty feet in front of me was a deer. I almost swerved off the road to avoid hitting it, screeching the brakes as I did, but my daughter still managed to keep on sleeping. With the eerie feeling that my dad was in the car with us, I couldn't help but think about the remarkable events that had occurred since he and I went to the 2009 D-Day celebration.

~~~

Betsy and I arrived at Saint-Malo, France, to catch the 8:00 a.m. ferry. to Jersey Island. When the ferry landed on Jersey Island, John Ovenden and his family met us. John had agreed to host us during our visit. John's wife had made us a reservation at a bed-and-breakfast overlooking one of the Jersey Island harbors, and John dropped us off there after a short ride from the ferry port.

"Bob, I suspect that you would like to get some sleep," he said. "But when you're ready, I have some research you can look at about the *PT-509* battle."

John Ovenden had discovered the wreckage of *PT-509*, while scuba diving a few hundred yards off the coast of Jersey Island. Others had found the wreakage before him and done nothing about it, but not John. He explored it in detail until he found a serial number on the engine. He backtracked this number to PT Boats, Inc. in the US, who put him in touch with some of the surviving members of PT Squadron 34. John then scheduled meetings with many of these men at a yearly PT boat reunion. He videotaped interviews of all of the men who knew about *PT-509's* battle and the subsequent rescue attempt. He put it all together into a documentary movie called *PT-509, The Last Patrol, Battle for the British Channel Islands.*

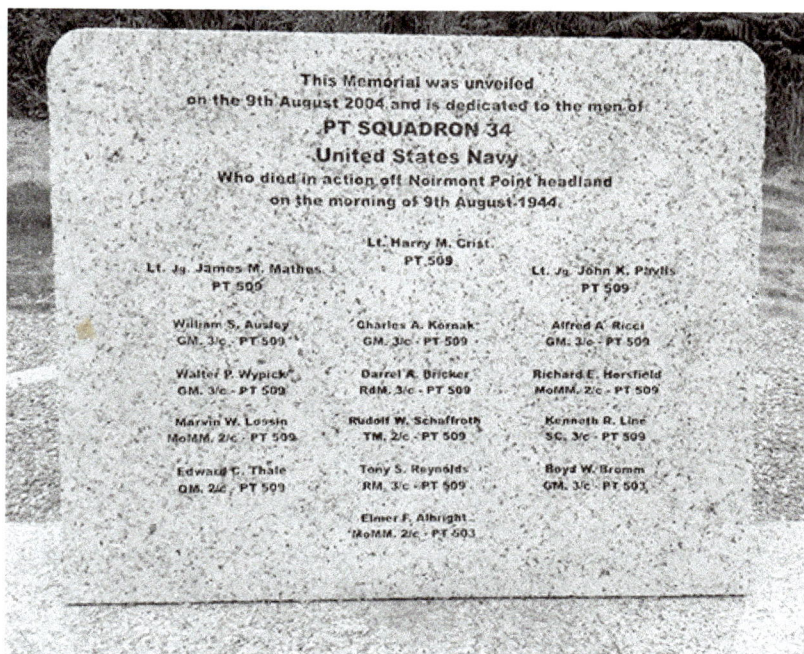

Figure 121. Memorial to RON 34 men who died in action off the coast of Noirmont Point, Jersey Island on August 9th 1944. Created and Dedicated by John Ovenden on August 9th 2004.
R. Sherertz collection.

By the time he finished his documentary, he had become extremely invested in my dad's Squadron 34. He subsequently created a memorial plaque that sat on a hill on the Jersey Island coast overlooking the site of the *PT-509* battle with the German boats and even named his two children after one of the men in my dad's squadron. He took us to see the memorial plaque and then to the cemetery where three of the men killed in the attack were buried (Darrel Bricker, Richard Horsfield, and Rudolph Schaffroth; August 15, 1944). Looking at their headstones, knowing that my father had given the order that sent these men to their deaths, gave me a gut-wrenching feeling that I could only imagine was a hundred times worse for him.

Figure 122. Jersey Island RON 34 graves. Darrel Bricker,
Richard Horsfield, Rudolph Schaffroth.
R. Sherertz collection.

After little more than a day our visit to Jersey Island was over and we drove back to Paris and arrived exhausted.

~~~

Today is D-Day, I thought to myself as I made the drive with my daughter Betsy from Paris to Lille, France. It was going to be an emotional roller-coaster ride for me. Betsy slept through most of the two hours, but I was so revved up that, though exhausted, I wouldn't have been able to sleep, even if I'd had the time. We arrived in Lille at the convention center and parked the car in the underground parking area. Then we found a map of Lille and some suggested destinations so that my daughter could explore the city, and I found the meeting organizers to let them know I had arrived.

Usually when I gave a talk, I experienced last-minute jitters as a normal course of affairs. But I was so steeped in thinking about my father that when they finally called my name to come up to the podium to speak, I was almost surprised. I methodically went through the slides I had prepared for this group of French physicians and nurses and then got to my last slide. It showed my father in his navy blue dress uniform, with a few other pictures relevant to France. I started to tell them about him.

"Sixty-eight years ago today, my father was the executive officer of PT Boat Squadron Thirty-Four, which guarded the western flank of the D-Day invasion force. Subsequently, his squadron patrolled the waters surrounding the German-occupied Channel Islands to make sure that the German boats did not disrupt supply lines to France. When Cherbourg was captured by the Allies, my father became the commanding officer of the Allied naval base established there. The images adjacent to my father's picture show the damage to the Cherbourg fort from German shelling, the only remains of *PT-509* found, one of the boats in Squadron 34 which sank in a battle off the coast of Jersey Island, and one more important thing. Last fall, sixty-seven years after D-Day, my father was awarded the French Legion of Honor Medal by French

Consulate General Pascal Le Deunff in Atlanta, Georgia. I would like to thank the French people and the French government for recognizing my father's efforts on their behalf."

Figure 123. Lt. Commander Sherertz with French Legion of Honor medal (left lapel) standing next to French Consulate General Pascal le Deunff.
R. Sherertz collection.

As I went through the slides, I could feel the emotion welling up inside, and I was barely able to get through the thank-you at the end. Unable to say anything further, all I could do was give a thumbs-up and a nod, signifying that I was finished. At first, dead silence ensued. Then applause began and increased until it became quite loud, with many people standing as they applauded. When the applause ceased, I looked over at the moderator of the session, as this would normally be the point where she would ask if there were any questions. She sat there, not speaking, with tears running down her face.

After a moment the other session moderator spoke, in a voice thick with emotion. "Thank you. We will take a short break before our next speaker."

I got down off the stage and stood there, collecting myself. On their way out for the break, the people from the audience walked by me, almost all of them with tears on their faces. A few tried to speak but couldn't. Several hugged me. The moderator who initially couldn't speak herself, came over and hugged me, too. I was totally unprepared for the depth of reverence still held by the French for the Allies coming to their aid. Fortunately, I had nothing further to do professionally that day, as I was emotionally spent.

~~~

The next morning, my daughter and I boarded a plane at Charles de Gaulle Airport to return to Charlotte, North Carolina. There my daughter and I parted ways, as she was heading back to Greensboro, and I got in my truck to go see my father. I arrived with great trepidation, not knowing what I would find, since the last time I had seen him he was almost unresponsive.

To my great and joyous surprise, I found him sitting up in the hospital bed that had been in the living room for the past few months, watching an old *Lawrence Welk* rerun and swaying to the music. I broke out laughing. As soon as he heard me, he beamed and called out my name. It came out clearly for the first time in months.

I took his hands in mine, and he didn't let go for almost fifteen minutes. While we held on to each other, I told him about our trip to Jersey Island and the French appreciation of his efforts. Tears came down his face and mine during the rest of the time I was there.

Finally, he looked at me and said, almost in a whisper. "Thank you for your time." Those were the last words he ever spoke to me.

~~~

Three weeks later, on June 28, 2012, my father died. As a World War II veteran he was eligible for burial with full military honors in Arlington Cemetery in Washington, DC. This possibility was discussed with my father prior to his death, but he was not interested. At his request he was cremated; half of his ashes were buried in Buchanan, Virginia next to my mother and half of his ashes were buried in Mint Hill, North Carolina near his surviving second wife.

Both funerals had a military honor guard with the Virginia group firing a nine-gun salute.

During both funerals I kept thinking about the last thing he said. Finally, it dawned on me. In the last three years of his life, he and I had talked more about things that mattered than in the other fifty-nine years I was alive. While I think he truly wanted our family to understand what he went through during World War II, our conversations and time together were more important.

Epilog

Honorably discharged from the Navy in February 1946, my father married my mother, Margaret McCulloch Sherertz, in February 1947. Later in 1947, after receiving one last call from John Bulkeley, he considered reenlisting and making the Navy his career. He saved correspondence indicating that the company he worked for had approved his going on a summer cruise with the Navy to try out the idea. A few months later, my mother gave birth to my older brother. He was born prematurely and died two days later. After that, my father decided that the right thing to do was to stay with his family, not go off on long cruises. That was the last he ever thought about going back to the Navy.

With his natural-born leadership abilities dating back to childhood, rapid wartime promotions, and a promise by a Congressional Medal of Honor winner of fast promotion after the war, one might have thought that my father would have ended up running a company or something equally ambitious. To many people's surprise, especially my mother's, he turned down all executive training offers. Instead, he took a job that required no employee supervision. I think he made this choice because during World War II, when he supervised men, they died.

He brought his family to Meadville, Pennsylvania, a

small bucolic town, that would have made a great Norman Rockwell painting for the cover of the *Saturday Evening Post*. There he was able to raise a family and keep his war demons at bay. Never again would he be Lieutenant Commander Sherertz, just Jack. Little did I know, he was one of the fortunate ones.

Only one time in my life did he mention wanting to be a physician. It was right after I finished my medical training when he told me how proud he was. When I asked him why he hadn't told me sooner, he said he hadn't wanted his past to influence my future. He was very emotional, and sometimes I couldn't understand what he was saying. In retrospect, I think talking about being a doctor briefly opened the floodgates on World War II.

One other price of World War II was losing interest in dancing. He would never say why. I think it had to do with memories of going dancing with his buddies the night before they died in the Pearl Harbor attack. As his memories softened late in life, he was finally able to start dancing again.

In the last three years of his life my father gave me an extraordinary gift. He shared the bottled-up pain from five years of World War II service. Since then I have spoken with many people having family members who served in World War II or another war. If their veteran experienced the horrors of war, most were unwilling to tell their story. One of the reasons I wrote my father's story was to help others understand the price soldiers pay. We must recognize that price—while they are alive—thank them for it, and remember it.

On my own behalf, I now fly a flag every day and thank each veteran I meet for their service. More importantly and with much pride, I can now close out my father's story comfortable in the knowledge that my children celebrate his sacrifice and will pass that knowledge on to their children. One final note, my grandchildren have become at least the fourth generation of Sherertz children to have "Frog Went a Courtin" sung to them by Sherertz men.

USS *Nevada*. My father always spoke with reverence about the USS *Nevada*, so I will tell you what happened to it after my father's last contact. After the USS *Nevada* finished shelling Cherbourg it went on to support the invasion of southern France near Marseilles at Cote d'Azur, then to support Iwo Jima operations via the Panama Canal, then finally to Okinawa, where it was hit by a Kamikaze plane. Postwar in 1946 it became part of a test fleet used to assess the impact of atomic bombs on naval vessels (Operation Crossroads). It survived five nuclear explosions. Then in July 1948 it was towed to a location 65 miles south of Pearl Harbor, where it took three attempts to finally sink her (latitude 20deg 58min north, longitude 159deg 17min west).

Figure 124. Lt. Commander Sherertz at World War II Memorial in Washington DC as part of NC Rotary Flight of Honor.

Figure 125-A. My son's tattoo.

Figure 125-B. My daughter's tattoo

Thanks for sharing, Dad. I miss you.

References

Books:

Pearl Harbor:

Stillwell, Paul, ed., *Air Raid:Pearl Harbor! Recollections of a Day of Infamy*, Annapolis: Naval Institute Press, 1981.

Wren, LCDR L. Peter and Sehe, Charles T., *Battle Born: The Unsinkable USS Nevada BB-36*, Bloomington: Xlibris Corporation, 2008.

Prange, Gordon W., Goldstein, Donald M., Dillon, Katherine V., *Dec. 7 1941: The Day the Japanese Attacked Pearl Harbor*, New York: Warner Books, 1989.

Lord, Walter, *Day of Infamy*, Pennsylvania State University: Bantam Books, 1957.

Toland, John, *Infamy, Pearl Harbor and its Aftermath*, Garden City: Doubleday & Co., Inc., 1982.

Toll, Ian W., *Pacific Crucible, War at Sea in the Pacific, 1941-1942*, New York: W. W. Norton & Company, 2012.

Barker, A.J., *Pearl Harbor, Ballantine's Illustrated Battle History of WWII, Book No.10*, New York: Ballantine Books, 1969.

van der Vat, John, *Pearl Harbor, Day of Infamy – An Illustrated History*, New York: Basic Books, 2001.

LaForte, Robert S. and Marcello, Ronald E., Ed,, *Remembering Pearl Harbor, EyeWitness Accounts by U.S. Military Men and Women*, Lanham: Rowman & Littlefield Publishers, 1991.

Shanks, Sandy, *The Bode Testament*, New York: Writer's Club Press, 2001.

van der Vat Dan, *The Pacific Campaign, The U.S.-Japanese Naval War 1941-1945*, New York: Simon & Schuster, 1991.

Victor, George, *The Pearl Harbor Myth, Rethinking the Unthinkable*, Dulles: Potomac Books, Inc., 2007.

Jasper, Joy Waldron, Delgado, James P., Adams, Jim, *The USS Arizona: The Ship, the Men, the Pearl Harbor Attack, and the Symbol that Aroused*

America New York City: St. Martin's Press, 2003.

Midway:

Prange, Gordon W., Goldstein, Donald M. and Dillon, Katherine V., *Miracle at Midway,* New York: Penguin Books, 1983.

Moore, Stephen L., *Pacific Payback, The Carrier Aviators Who Avenged Pearl Harbor at the Battle of Midway,* New York: Penguin Group, 2014.

Astor, Gerald, *Semper Fi in the Sky, The Marine Air Battles of World War II,* New York: Presidio Press, 2005.

Coale, Lieutenant Commander Griffith Baily, USNR, *Victory at Midway,* New York: Farrar & Rinehart, 1944.

D-Day:

Time, *D-Day, 24 Hours That Saved the World, Time Magazine 70th Anniversary Tribute,* New York City: Time Magazine, 2014.

Ambrose, Stephen E., *D-Day, June 6, 1944: The Climactic Battle of World War II,* New York: Simon & Schuster, 2001.

Beevor, Antony, *D-Day, The Battle for Normandy,* New York: Penguin Group, 2009.

Astor, Gerald, *June 6,1944, The Voices of D-Day,* New York: Dell Publishing, 1994.

Ford, Ken, Zaloga, Stephen J., *Overlord, The Illustrated History of the D-Day Landings,* Oxford: Osprey Publishing, 2011.

Penrose, Jane, Ed., *The D-Day Companion, Leading Historians Explore History's Greatest Amphibious Assault,* Oxford, Osprey Publishing, 2009.

Ross, John, *The Forecast for D-Day, And the Weatherman behind Ike's Greatest Gamble,* Guilford: Lyons Press, 2014.

Ryan, Cornelius, *The Longest Day, The D-Day 70th Anniversary Collector's Edition,* New York: Barron's Educational Series, 2014.

Balkoski, Joseph, *Utah Beach, The Amphibious Landing and Airborne Operations on D-Day, June 6, 1944,* Mechanicsburg: Stackpole Books, 2006.

PT boats:

Bachman, Bruce M., *An Honorable Profession, The Life and Times of One of America's Most Able Seamen: Rear Adm. John Duncan Bulkeley, USN,* New

York: Vantage Press, 1985.

Bulkley, Robert J., Jr, *At Close Quarters: PT Boats in the United States Navy*, Annapolis: Naval Institute Press, 1962.

Breuer, William, *Devil Boats: The PT War Against Japan*, New York: Presidio Press, 1987.

Nelson, Curtis L., *Hunters in the Shallows*, Dulles: Potomac Books, Inc., 1998.

Smith, George W., *MacArthur's Escape, John "Wild Man" Bulkeley and the Rescue of an American Hero*, Minneapolis: Zenith Press, 2005.

The Bureau of Ships, *Know Your PT Boat*, Annapolis: US Navy Technical Publication No. 9 1945.

Schuster, Russell E., Allen, Stanley, and Bosley, Sheldon, with a foreword by Jack Sherertz, *Motor Torpedo Boat Squadron 34 History*, Germantown: PT Boats, Inc., 2001.

Restricted Operating Manual, Packard Marine Engine 4M-2500 Types W7, W8, W9, W10, Packard Motor Car Company Marine Engine Division, 1943.

Farley, Edward I., *PT Patrol*, Whitefish: Literary Licensing, LLC, 2011.

US Navy, *Motor Torpedo Boat Manual*, Washington: US Government, 1943.

Breuer, William B., *Sea Wolf: The Daring Exploits of Navy Legend John D. Bulkeley*, New York: Presidio Press, 1998.

Ovenden, John and Shayer, David, *Shipwrecks of the Channel Islands*, Guernsey: Underwater Video Services, 2002.

Hoagland, Lt. Comdr. Edgar D. (Ret.), The *Sea Hawks, with the PT Boats at War, A Memoir*, New York: Presidio Press, 1999.

White, W. L., *They Were Expendable*, Annapolis: Naval Institute Press, 1998.

Navy:

Bunkley, J. W., *Military and Naval Recognition Book*, New York: D. Van Nostrand Company, Inc., 1941.

Lovette, Leland P., *Naval Customs, Traditions and Usage*, Annapolis: Naval Institute Press, 1939.

Ageton, Commander Arthur, A., *Naval Officer's Guide*, New York:

Whittlesey House, 1943.

US Navy, *Naval Leadership, with Some Hints for Junior Officers and Others*, Annapolis: Naval Institute Press, 1939.

Conger, Elizabeth M., *Ships of the Fleet*, New York: Henry Holt and Company, 1946.

Fahey, James C., *The Ships and Aircraft of the U.S. Fleet*, New York: Gemsco, 1939.

World War II:

Overy, Richard, Ed., Foreword by Tom Brokaw, *The New York Times Complete World War II, 1939-1945, The Coverage from the Battlefields to the Home Front with Access to 98,367 Articles*, New York: Black Dog & Leventhal Publishers, 2013.

John Ford:

Stoehr, Kevin L. and Connolly, Michael C., *John Ford in Focus, Essays on the Filmmaker's Life and Work*, Jefferson: McFarland, 2007.

Ford, Dan, *Pappy: The Life of John Ford*, Upper Saddle River: Prentice Hall, 1979.

Midshipmen's School Yearbook:

Midshipmen, *Side Boy The Second Class, United States Naval Reserve, Midshipmen's School, USS Prairie State*, New York: US Navy, February 1941.

Movies or Videos Used as References:

Ford, John, Director, *The Battle of Midway*, War Activities Committee, 1942.

Ovenden, John and Haslam, Paul, Directors, *PT- 509, The Last Patrol, Battle for the British Channel Islands*, Jersey Island: Underwater Video Services 01534 484599, 2001. Documentary film that describes a battle between US RON 34 PT boats and the German navy on August 9, 1944, near Jersey Island.

Harding Drane discussing Pearl Harbor attack, Cuyahoga County Public Library website, http://www.worldcat.org/title/walter-harding-hardy

-drane-pearl-harbor-survivor/oclc/53813666&referer=brief_results, Harding Drane went through V-7 officer training with Jack Sherertz and was on USS *Nevada* with Jack Sherertz at Pearl Harbor.

Web references:

"Burial at Sea," http://www.history.navy.mil/faqs/faq85-1.htm
"Utah Beach to Cherbourg," War Department Document published in 1948, http://www.history.army.mil/books/wwii/utah/utah.htm
"World War II US Convoy Formation," last updated 12/8/14, convoycu49-1944.com
"World War II US Convoy Routing," World War II-- file:///Users/sherertz/Desktop/Convoy%20and%20Routing%20-%20Administrative%20History%20of%20World%20War%20II.html
"D-Day Radio broadcasts," *https://archive.org/details/Complete_Broadcast_Day_D-Day*
"Hawaiian Island chain up to Midway," http://radiojerry.com/frigate/hawaiimp.jpg
"US Marines at Midway," http://www.au.af.mil/au/awc/awcgate/usmchist/midway.txt
"Midway Ships and Aircraft 1942," http://midway1942.org/ships/usn_dd_porter.shtml
"Radio Reports Japan's Attack on Pearl Harbor," http://www.modestoradiomuseum.org/radio%20reports%20pearl.html
"Operation Neptune," D-Day US Navy assault forces, http://www.history.navy.mil/library/online/comnaveu/comnaveu-6.htm
"Pearl Harbor Japanese Attack Chronology," http://www.navsource.org/Naval/logs.htm
"PT Boat, Humorous Introduction," World War II US Bureau of Ships—http://www.pt171.org/PT171/sting/
"World War II radio broadcasts: Pearl Harbor - https://library.umkc.edu/spec-col/ww2/pearlharbor/radio.htm, D-Day - http://www.wwiifoundation.org/students/real-time-radio-broadcasts-from-d-day-june-6-1944/?gclid=CjwKEAjw4dm6BRCQhtzl6Z6N4i0SJADFPu1n0MZyMgeCUIE66-WtY5kDjOzI9hYOkeQV1UChAzOjOhoC5hPw_wcB , Comprehensive

list of World War II radiobroadcasts (must be ordered) - http://
www.archives.gov/research/military/ww2/sound-
recordings.html#41 , search by date - http://
www.radiogoldindex.com/frame1.html (must be ordered,
recommended by Greg Bell of GregBellMedia.com).
"SB2U Vindicator," http://www.vought.org/products/html/
sb2u.html
"*USS Nevada Torpedo and Bomb Damage*," file:///Users/sherertz/
Desktop/BB-36%20USS%20Nevada%20Pearl%20Harbor%20Attack%
20Damage%20Report.webarchive

Appendices

Appendix 1. Herbert Jackson Sherertz WWII Chronology

Assignments	Dates	Location	Rank
USS *Arkansas*	9/30/40 – 10/25/40	NYC	Midshipman
USS *Illinois*	11/22/40 – 2/26/41	NYC	Midshipman
USS *Nevada*	4/4/41 – 12/17/41	Pearl Harbor	Ensign
MTB RON 1	12/17/41 – 5/29/42	Pearl Harbor	Ensign
MTB RON 1	5/30/42 – 10/15/43	Midway	Ens, Lt Jg, Lt
MTBSTC	11/13/43 – 12/14/43	Melville, RI	Lieutenant
MTB RON 34	12/14/43 – 3/13/45	NYC, D-Day	Lt, Lt Cmdr
MTBSTC	3/13/45 – 7/6/45	Melville, RI	Lt Cmdr
ComMTB Pac	7/6/45 – 8/15/45	Philippines	Lt Cmdr

Appendix 2. V-7 Training Cruise, Colon to Panama City, Partial Roster Section 1, Division 6

Number	Apprentice Seaman Name	
316	O'Leary, P.V.	Squad Leader (1st)
317	Rowland, D.C.	
318	Whitehead, E.H.	
319	Wengor, P.B.	
320	Sherertz, H.J.	
321	Sturkey, C.M.	
322	Twomey, B.F.	
323	Shannon, E.F.	Squad Leader (2nd)
324	Griese, W.C.	
325	Saul, J.P.	
326	Taylor, F.E.	
327	Skinzich, M.C.	
328	Sislo, W.A.	
329	Buchmann, R.R.	
330	Yates, Z.C.	Squad Leader (3rd)
331	Weber, G.E.	
332	Stromberg, K.R.	
333	Hutzborg, J.M.	
334	Brown, D.A.K.	
335	Bachrach, A.	
336	Anderson, J.W.	

Appendix 3. Jack Sherertz Pearl Harbor Ensign friends; most graduated with The Second Class, United States Naval Reserve, Midshipmen's School, USS *Prairie State*, New York, New York.

		Assigned	Ship	
Thomas R. Jones	Ensign USNR	2/24/41	USS Arizona	Died
Andrew C. Uhrenholdt	Ensign USNR	2/24/41	USS Arizona	Died
Jerome H. Garfield	Ensign USNR	2/24/41	USS Arizona	Survived
Robert S. Booth	Ensign USNR	11/14/40	USS Arizona	Died
John L. Mote	Ensign USNR	2/24/41	USS Nevada	Survived
Walter H. Drane	Ensign USNR	2/24/41	USS Nevada	Survived
Herbert J. Sherertz	Ensign USNR	2/24/41	USS Nevada	Survived
Harold J. Christopher	Ensign USNR	8/5/41	USS Nevada	Died
Frederick C. Davis	Ensign USNR	9/4/41	USS Nevada	Died
Joseph K. Taussig, Jr	Ensign USN	3/2/41	USS Nevada	Survived
Charles M. Sterns, Jr.	Ensign USNR	2/24/41	USS Oklahoma	Died

Appendix 4. Officers attached to the USS *Nevada* as of December 7th 1941. Number after name = Naval Academy graduation date. On Board (OB) = whether present at start of PH attack. Source unclear, Jack Sherertz's personal records.

Name	On Board?	Rank	Start Date	Duty	K.I.A.
Francis W. Scanland '09	NOB	Captain	6/4/41	Commanding officer	
Harry L. Thompson	NOB	Commander	7/25/41	Executive officer	
George C. Miller '21	NOB	Lt-Comdr	7/12/40	1st Lt. & D.C.O.	
Armand J. Robertson '22	NOB	Lieut-Comdr	6/12/40	Gunnery officer	
William L. Freseman '22	NOB	Lieut-Cmdr	9/18/41	Navigator	
George I. Fee '25	NOB	Lieut-Cmdr	6/8/39	Engineer officer	
Francis J. Thomas '25	OB	Lieut-Cmdr USNR	6/22/41	Asst. 1st Lt & D.C.O.	
Arthur R. Quinn '25		Lieutenant	1/13/41	Asst. Gunnery officer	
Robert DeC. Baker '26		Lieutenant	8/29/41	Main Battery officer	
James C. Clarkson '30		Lieutenant	6/16/39	Asst. Engineer officer	
Lawrence E. Ruff '30	OB	Lieutenant	2/17/41	Comm. off. & ship sec.	
Benjamin T. Waldo		Lieutenant USNR			
John O. Spear '32		Lieutenant	9/8/41		
James D. Furguson '33 NOB		Lieutenant	6/1/39	Air defense OFC	
Robert B. Conaughty		Lieutenant (jg) USNR	1/20/41	Radio officer	
Marvin B. Miller		Lieutenant (jg)			
Harrison H. Holton '38		Lieutenant (jg)	6/29/38	3rd Division officer	
Lionel T. McQuiston '38		Lieutenant (jg)	3/2/41	Senior officer	
George W. Scott '39		Ensign	7/15/39		
Harry A. Seymour '39 OB		Ensign	7/15/39	6th Division officer	
Charles W. Jenkins '39 OB		Ensign	7/15/39	3rd Division J.O.	
Ernest H. Dunlap Jr '39OB		Ensign	7/15/39		
George E. Thode		Ensign USNR	11/10/39	5th Division officer	
William R. Leonard Jr		Ensign USNR		Junior aviator	
Luther R. Johnson		Ensign USNR			
William R. Boehm '40		Ensign	7/15/40	E Division J.O.	
Joseph R. Treanor '40		Ensign	7/12/40	F Division J.O.	
Richard J. Nesbitt '40	OB	Ensign	7/12/40	4th Division J.O.	
Ward W. Witter '40		Ensign	7/12/40	6th Division J.O.	
Raymond E. Hill '40	OB	Ensign	7/12/40	2nd Division officer	
Charles M. Wood '40		Ensign	7/12/40	Signal officer	
Thomas H. Taylor '40	OB	Ensign	7/2/40	6th Division J.O.	
Allen P. Cook Jr '40		Ensign	7/15/40	B Division officer	
Robert E. Jeffrey '40		Ensign	7/2/40	1st Division officer	
James G. Egan '40		Ensign	7/15/40	A Division officer	
Frederick C. Davis		Ensign USNR		Junior aviator	
John C. Borden		Ensign USNR	12/1/40	Asst. Signal officer	
Daniel E. Elmore	NOB	Ensign USNR (Detroit)	12/1/40	5th Division J.O.	
Delar V. Van Sand		Ensign USNR	12/1/40	B Division J.O.	
Joel H. Zaritsky		Ensign USNR			
Julian W. Bailey		Ensign USNR	12/29/40		
Barton L. Fischer		Ensign USNR	12/29/40		
John W. Hollander		Ensign USNR	12/29/40	C.W.O.	
Lynn F. Barry '41		Ensign	3/2/41		
John L. Landreth '41	OB	Ensign	3/2/41	6th Division J.O.	
Charles J. Merdinger '41		Ensign	3/2/41	F Division J.O.	
Joseph K. Taussig '41	OB	Ensign	3/2/41		
Robert E. Thomas Jr '41	OB	Ensign	3/2/41	6th Division J.O.	
John L. Mote		Ensign USNR	2/24/41	A Division J.O.	
Walter H. Drane		Ensign USNR	2/24/41		
Herbert J. Sherertz	NOB	Ensign USNR	2/24/41		
Charles H. Grainger		Ensign USNR	4/26/41		
Charles F. Berg		Ensign USNR	4/26/41	2nd Division J.O.	

Name	On Board?	Rank	Start Date	Duty	K.I.A.
Robert J. Rowlands		Ensign USNR	4/26/41		
William M. Filmer		Ensign USNR	4/26/41		
Allen J. Huttenberg	OB	Ensign USNR	4/26/41		
Lon L. Laymon	OB	Ensign USNR	4/26/41	6th Division J.O.	
John J. Ryan		Ensign USNR			
William W. West Jr		Ensign USNR	7/18/41	C.W.O.	
Raymond J. Wise Jr		Ensign USNR	7/18/41		
Wyman G. Smith Jr		Ensign USNR	7/18/41		
Harold J. Christopher	OB	Ensign USNR	8/5/41	Yes	
Samuel C. Jackson		Ensign USNR		Junior Aviator	
Francis T. Baldy		Ensign USNR	10/25/41		
Phelan C. Hawn		Ensign USNR	10/25/41		
Robert E. Lemmon		Ensign USNR	10/25/41		
Robert D. Dewar		Ensign USNR	10/25/41		
Bruce B. Edwards		Ensign USNR	10/25/41		
Robert B. Hegler		Ensign USNR	10/25/41		
Loren K. Hoff		Ensign USNR	10/25/41		
Francis J. Sullivan		Ensign USNR	10/25/41		
John W. Morrison		Ensign USNR	10/25/41		
Edward J. Goodbody		Commander (MC)	3/2/41	Senior Medical officer	
Jerome F. Zobel		Lieut. (jg) (MC) USNR	3/2/41	Junior Medical officer	
Robert A. Freyling		Lieut. (jg) (MC)	6/21/41	Junior Medical officer	
Curtiss W. Schantz		Lieutenant (DC)		Dental officer	
William C. Colbert		Commander (SC)	7/20/40	Supply officer	
William A. Rapp		Ensign (SC) USNR	8/40	Asst. Supply officer	
Benajah L. Rainer		Ensign (SC) USNR		Asst. Supply officer	
Raymond B. Drinan		Lieut-Comdr (ChC)		Chaplain	
Walter Asmuth Jr		Captain USMC	6/22/41	Comdr Marine detachment	
Stoddard G. Cortyelyou		2nd Lt. USMC	7/10/41	Junior Marine officer	
Paul T. Johnston	NOB	2nd Lt. USMC	11/9/41	Junior Marine officer	
Edwin J. Hill	OB	Chief Boatswain			
James L. Treadwell		Gunner	3/23/37		
Joshua H. Garrett		Chief Machinist	2/1/41	M Division J.O.	
Donald K. Ross	OB	Machinist		A Division J.O.	
Reynolds F. Bess		Electrician	10/6/40		
Eugene L. Conant		Carpenter	5/24/41		
John W. Cooper	OB	Pay Clerk	12/11/39		
Albelbert W. Dutcher		Pay Clerk	6/9/40		

Appendix 5. USS *Nevada* Casualties During Pearl Harbor Attack

ANDERSON, Arnold Leo	S1c	USN
AQUINO, Zoilo	MATT1c	USN
BINGHAM, James Robert	S2c	USN
BLEDSOE, Herman	MATT2c	USN
BRIGGS, Lyle Lee	EM2c	USN
BRITTON, Thomas Alonzo	CPL	USMC
CHRISTOPHER, Harold Jensen	ENS	USNR
COOK, Joseph William	GM3c	USN
CORBIN, Leon John	GM1c	USN
COTNER, Leo Paul	S2c	USN
DAVIS, Frederick Curtis	ENS	USNR
DUKES, Lonnie William	S1c	USN
ECHOLS, Edward Wesley	COX	USN
EDWARDS, Harry Lee	S1c	USN

FADDIS, George Leon	GM3c	USN
FUGATE, Kay Ivan	S1c	USN
GANTNER, Samuel Merritt	BM2c	USN
GILES, Thomas Robert	EM3c	USN
GOETSCH, Herman August	S1c	USN
GULLACHSON, Arthur K.	S2c	USN
HALLMARK, Johnnie W.	S1c	USN
HARKER, Charles Ward	FC3c	USN
HEATH, Francis Colston	PFC	USMC
HEIM, Gerald Leroy	S2c	USN
HILL, Edwin Joseph	CWO	USN
HUBNER, Edgar E.	S1c	USN
IRISH, Robert Clement	S2c	USN
JOHNSON, Flavous B.M.	GM3c	USN
KING, Orvell Vaniel Jr.	PFC	USMC
LAMONS, Kenneth Taft	BM2c	USN
LIPE, Wilbur Thomas	S2c	USN
LUNSFORD, Jack Leon	PFC	USMC
LUNTTA, John Kallervo	S1c	USN
MAFNAS, Andres Franquez	MATT1c	USN
MARTIN, Dale Lewis	SC1c	USN
MAYFIELD, Frazier	MATT1c	USN
McGHEE, Lester Fred	S1c	USN
McGUCKIN, Edward L.	S1c	USN
MORRISSEY, Edward Francis	PFC	USMC
NEUENDORF, William F. Jr.	S1c	USN
NORVELLE, Alwyn Berry	CSKA	USN
PATTERSON, Elmer Marvin	CK2c	USN
PECK, Eugene Edward	S2c	USN
ROBISON, Mark Clifton	MATT1c	USN
RONNING, Emil Oliver	COX	USN
RUSHFORD, Harvey George	S2c	USN
SCHWARTING, Herbert C.	S1c	USN
SHAUM, Donald Robert	S1c	USN
SMITH, Keith Vodden	PVT	USMCR
SOLAR, Adolfo	BM1c	USN
SPEAR, Herman Alder	S1c	USN
SPENCER, Delbert James	S1c	USN
STEMBROSKY, George Joseph	S1c	USN
STRICKLAND, Charles E.	S1c	USN
THUNHORST, Lee Vernon	S2c	USN
TRUJILLO, Richard Ignacio	PFC	USMC
WALTON, Ivan Irwin	COX	USN

Appendix 6. MTB RON 1 April 15, 1942 Boat Crews. Parentheses indicate boat locations after the Battle of Midway.

Squadron Commander Lt McKellar

PT-20 (Midway)		*PT-21 (New Guinea, 7/42)*		*PT-22 (Alaska, 7/42)*	
Ensign	Rennell	Ensign	Preston	Lt (jg)	Parker
Ensign	Callaghan	Ensign	Faulkner		
CGM	Ross	GM.3c	Christiansen	GM.2c	Blozis
SM.3c	Stewart	QM.2c	Valentine	QM.2c	Lunday
MM.1c	Charbonneau	RM.3c	Scribner	Sea.2c	Watson (radio)
MM.2c	Waite	CMM	Hunter	MM.1c	Martin
A.S.	Sebolt	CMM2c	Leonard	MM.1c	Featherling
SC3	Adams	A.S.	Zicarelli	MM.2c	Hayes
Sea1c	King (ord.)	A.S.	Collins (cook)	A.S.	Westburg (cook)

PT-23 (New Guinea, 7/42)

Ensign	Howes
Ensign	Morrison
GM.2c	Leonard, G.H.
QM.3c	Morrison
RM.3c	Langley
CMM	Tiller
MM.1c	Strombert
MM.2c	Sitton
A.S.	Slagle (cook)
A.S.	Sanko (ord.)

PT-24 (Alaska, 7/42)

Ensign	Williams
TM.2c	Mott
Sea1c	Anderson (Q.M.)
RM.1c	Stockdale
MM.2c	Hubbard
MM.2c	Hurley
F.1c	Rush
A.S.	Simonson (cook)

PT-25 (New Guinea, 7/42)

Lt (jg)	Rice
Ensign	J Mote
CGM	Kruger
QM.1c	Fairweather
RM.2c	Walker
MM.1c	Burke
MM.2c	Penny
A.S.	Rothery
SC.2c	Robinson

PT-26 (New Guinea, 7/42)

Ensign	Ewell
Ensign	Jones
TM.1c	Barnes
QM.3c	Gardner
RM.3c	Stanfield
MM.2c	Blake
MM.2c	Franklin
A.S.	Tolley
Sea1c	Mills (cook)

PT-27 (Alaska, 7/42)

Ensign	Erikson
Ensign	Spoerer
TM.1c	Parks
QM.2c	Cheney
RM.3c	Hoover
MM.1c	Jester
MM.2c	Ekeman
F.1c	Toler
SC.2c	Otto

PT-28 (Alaska, 7/42)

Ensign	Williamson
Sea.1c	Hiatt (ord.)
QM.3c	Toomey
RM.1c	Landrum
MM.1c	Miastkowski
MM.2c	Lowell
F.3c	Miastkowski
SC.2c	Du Pont

PT-29 (Midway)

Lt (jg)	Matteson
Ensign	Sherertz
GM.3c	Gifford
CQM	Ritter
Sea2c	Hess (radio)
MM.2c	Capps
MM.2c	Hagen
F.1c	Wilson
SC.1c	Cunningham

PT-30 (Midway)

Ensign	Dupzyk
Sea1c	Stanton (ord.)
CQM	Orrell
RM.3c	Hurd
MM.1c	Brown
MM.2c	Britt
A.S.	Serwon
SC.2c	Daniel

PT-42 (Midway)

Ensign	Bryant
Ensign	Blackstock
TM.3c	Bebeau
QM.2c	Gebhardt
A.S.	Stewart (radio)
MM.1c	Sawicki
MM.2c	Miller
A.S.	Sparrow
SC.1c	Tajkowski

Appendix 7. RON 1 (*PTs 20, 29, 30, 42*) After Battle of Midway

Figure 126. Group picture 6 months after Midway.

(continues on next page)

421

Figure 127. RON 1 Officers at Midway. Lt. Parker, Lt. Sherertz (arrow), Lt. Chase, Lt. Bryant, Ens Godik, Ens Bynum, Ens Saltsman (Not present for picture).
Jack Sherertz collection.

Figure 128. Lt (jg) Sherertz's (CO, arrow), Crew – 1st PT boat Command, PT-29, after Battle of Midway Ens. Saltsman (right), Ritter, C.Q.M., Hagan, M.O.M.M. 1/c, Haas, R.M. 3/c, Hill, G.M. 3/c, Forrman, Sea 1/c, Whitehurst F 1/c, Floyd, F 1/c.
Jack Sherertz collection.

Figure 129. A,B,C. PTs *20, 30, 42* Officers and Crew at Midway. Jack Sherertz collection

12/31/43-7/16/44   CO – Lt. A.H. Harris, XO Lt. H.J. Sherertz
7/16/44-12/44   CO – Lt. H.J. Sherertz, XO Lt. J.J. Daniels

PT-498, *Jolly Roger*		PT-499, *Miss Kate*		PT-500, *Stratus*	
Lt. (jg)	W.S. Squire, CO	Lt. (jg)	H.G. Fraser, CO	Lt.	D.S. Kennedy, CO
Ens.	W.R. DeYoung, XO	Lt. (jg)	L.E. Pierce, XO	Ens	P.C. Washburn, XO
GM.2c	W.C. Peters	Lt. (jg)	R.H. Stuhler, 3rd O	Ens	S.D. Allen, 3rd O
TM.2c	L. Aulick	TM.2c	W.N. Kruger	QM.2c	K.M. Short
MM.2c	D.A. DeLapp	GM.2c	W.G. Schojain	MM.2c	R.L. Cook
MM.2c	P.F. Duff	QM.2c	R.G. Center	MM.2c	H.C. Weber
MM.2c	A.G. Anderson	MM.2c	L.N. Nearhood	MM.2c	A.M. Smith
SC.2c	R.P. Carter	MM.2c	G. T. Cummings	GM.2c	L.H. Leary
TM.2c	R.S. Bass	MM.2c	L.D. Wile	GM.3c	F. Reedy
RM.2c	W.R. Finn	RM.3c	R.M. Carlson	SC.2c	G.T. Pappas
GM.2c	P.J. Uhrich	SC.2c	J.M. Austin	TM.3c	H.S. Danneker, Jr
GM.2c	G.N. Christly	GM.3c	W.H. Barnes	RDM.3c	M. Leshock
QM.2c	W.D. Fox	RDM.3c	E.E. Pleau	RDM.3c	R.J. Morgan
SC.2c	W.R. Hamilton	S.1c	J.H. Harris	S.1c	P.J. Showalskis
S.1c	D.B. Vaage	RM.3c	R.J. Burns	RM.2c	F.J. Geppi
TM.3c	W.H. Pierce	GM.3c	W.J. Hyatt	F.1c	C.F. Henschen
F.1c	M.L. Patterson				
S.1c	F.P. O'Heaney				
S.2c	W. Kerecz				

PT-501, *none*		PT-502, *Idiot's Delight*		PT-503, *What Next*	
Lt. Ball, CO		Lt. (jg) R.P. Cooper, CO		Lt. J.A. Doherty, CO	
Ens.(?) Cooper, XO		Ens. R.L. Baker, XO		Ens. R.E. Schuster, XO	
GM.2c	F.E. Prosser	GM.2c	G.E. Marsden	Ens. F.H. Koenen, 3rd O	
GM.3c	D.W. Smith	GM.3c	C.W. Thomas	GM.3c	B.W. Brumm, KIA
GM.3c	MR. Parker	GM.3c	R.C. Ranieri	GM.3c	R.L. Biele
QM.2c	M.A. Smith	TM.2c	G.F. McClellan	TM.3c	J.M. Easterly
RM.3c	R.S. Fleming	MM.2c	L.G. Stephens	RM.2c	D.E. Fisher
MM.2c	C.C. Roby	F.1c	D.E. Carr	SC.3c	A.V. Duquette
MM.2c	J.L. Fadley	F.1c	C.P. Peterson	MM.2c	E.F. Albright, KIA
MM.2c	J.C. Suehla	SC.2c	H.M. Tyson	MM.2c	R.R. Allbee
RDM.3c	R.F. Noyer	RM.3c	P.H. Smith	RDM.3c	E.S. Kramer
RDM.3c	A.A. Markman	RDM.3c	M.L. Bailey	RDM.3c	C.O. Faucher
SC.1c	H.L. Blachler	RDM.3c	T.J. Boyle	MM.2c	P.B. Peppel
S.1c	W.J. Schmidt	QM.2c	E.A. Reichart	QM.3c	A.L. Lang
BKR.2c	F.R. Miles	S.1c	O.C. Solomon	GM.1c	E. Sullivan
TM.2c	L.K. Dixon	GM.3c	F.F. Arlak	GM.2c	J.L. Roberts
				MM.1c	P.C. Shadford
				MM.2c	J. Listar
				MM.2c	H.A. Heisler

PT-504, *Nasty Bastard*		PT-505, *Diana*		PT-506, *none*	
Lt. H.B. Sherwood, CO		Lt. W.C. Godfrey, CO		Lt. (jg) R.A. Bretell, CO	
Lt. H.W. Aldridge, CO		Ens. R.W. Hadley, Jr,		XO Ens. W.D. Surgeon, XO	
Ens. M.J. Sharkley, XO		Ens. A.R. Kuesel, 3rd off.		QM.2c	C.W. Arnold
QM.2c	R.W. Gretter	GM.1c	H.N. Miller, Jr	TM.2c	N.J. Nichter
RM.2c	G.A. Loehner	GM.2c	R.W. Wind	RM.2c	W.B. Gross
TM.2c	M.S. Carlson	TM.2c	M.H. Waldrons	MM.2c	C.F. Donahue
MM.2c	W.H. Aulish	QM.2c	R.L. Howard	MM.2c	R.A. Grant
MM.2c	L. Lapham	RM.2c	J. Frank	MM.2c	T.J. Hentges
MM.2c	E.F. Martin	MM.2c	W.E. Boling	GM.3c	M.E. Martin
GM.3c	J.B. O'Dell	MM.2c	W.W. Olson	SC.2c	E.H. Fenton
GM.3c	J.C. Puchalaky	MM.2c	C. Van Sickle, Jr	RM.3c	E.W. Burgelin
RDM.3c	C.M. Ahlers	GM.3c	W.R. Dunn	GM.1c	FH. Zeso

PT-504, *Nasty Bastard*
SC.2c   J.M. Archer
S.1c   F.D. Halahan
MM.3c   J.D. Haien
F.1c   J.H. Butler

PT-505, *Diana*
RM.3c   M.L. Kreeger
SC.2c   W.A. Cole
S.2c   E.L. Mowatt
F.1c   A.R. Hanson
S.1c   R. Mel
S.1c   J. Varga

PT-506, *none (continued)*
Cox.3c   P.C. Cayer
GM.3c   G.M. Smith
F.1c   R.L. Dory

PT-507, *Heminway Hotel*
Ens. B.T. Heminway, CO
Ens. J.B. Davis, XO
QM.3c   R.F. Hatmaker
RM.2c   J.C. O'Leary
GM.2c   R.F. Nieman
MM.2c   G.A. Fairchild
MM.2c   F.A. Johnson
MM.2c   O.A. Tucker
GM.3c   S.B. Bosley
GM.1c   D.L. Neves
TM.2c   D.R. Wilkinson
GM.3c   D.J. Lodato
GM.3c   C.W. Thomas
SC.2c   N.C. Hoyt
RDM.3c   C.E. Busse
GM.3c   J.S. Aldridge
Lt Crist (initial CO)

PT-508, *Mairsey Doats*
Lt. C.R. Whorton, CO
Lt. H.M. Aldridge, X
Lt. J.F. Queeney, 3rd
Lt. A. Reffler, Exec
GM.3c   M.J. Cunningham
QM.3c   R.A. Fitzpatrick
SC.2c   R.H. Jones
GM.3c   J.W. Permenter
TM.2c   J.A. Zagrocki
GM.2c   E.J. Tross
RDM.3c   S.E. Gilbert
RM.2c   A.P. Conte
QM.3c   S. Kaslow
MM.2c   A.E. Thompson
MM.2c   E.M. Hagan
MM.2c   C. Griffin
F.1c   S.J. Barkott
S.2c   R.A. Poling

PT-509, *Sassy Sue*
Lt. R.W. Netterstrom, CO
Ens. R.E. Schuster, XO
GM.3c   W.S. Ausley, KIA
RDM.3c   D.A. Bricker, KIA
MM.2c   R.E. Horsfield, KIA
GM.3c   C.A. Kornak, MIA
SC.3c   K.R. Line, MIA
MM.2c   M.W. Lossin, MIA
RDM.3c   J.L. Page, POW
RM.3c   T.S. Reynolds, MIA
GM.3c   A.A. Ricci, MIA
TM.2c   R.W. Schaffroth, KIA
QM.2c   E.C. Thale, MIA
GM.3c   W.P. Wypick, KIA
Lt H.M. Crist, CO, MIA
Lt (jg) J.K. Pavlis, XO, MIA
Lt (jg) J.M. Mathes, 3rd O, MIA

www.ingramcontent.com/pod-product-compliance
Lightning Source LLC
Chambersburg PA
CBHW071532200326
41519CB00021BB/6455